Notes on Biological Processes in Environmental Engineering

Varenyam Achal

Notes on Biological Processes in Environmental Engineering

Springer

Varenyam Achal
Department of Environmental Science
and Engineering
Guangdong Technion—Israel Institute
of Technology
Shantou, China

ISBN 978-981-95-1609-4 ISBN 978-981-95-1610-0 (eBook)
https://doi.org/10.1007/978-981-95-1610-0

This Springer imprint is published by the registered company Springer Nature Singapore Pte Ltd.
The registered company address is: 152 Beach Road, #21-01/04 Gateway East, Singapore 189721, Singapore

Preface

In a world where clean water and sustainable practices are increasingly vital, the ability to harness biological processes through engineering is more important than ever. *Notes on Biological Processes in Environmental Engineering* reflects my deep commitment to uniting the intricate science of microbiology with the practical demands of engineering. This book is carefully designed for engineering students who are eager to make a meaningful difference in wastewater treatment and environmental protection.

The content is organized into nine chapters that journey from the basics of microbiology to advanced treatment strategies. Beginning with fundamental concepts in "The Basic Microbiology: An Engineer's Perspective" and progressing through topics such as stoichiometry, bacterial energetics, metabolic reactions, microbial kinetics, and bioreactor operations, the book builds a cohesive framework essential for designing efficient wastewater treatment systems. Later chapters focus on the activated sludge process and advanced treatment methods, highlighting the integrated role of biological processes in safeguarding environmental quality.

I have intentionally employed a clear and accessible writing style, complemented by calculation-based problems and step-by-step solutions. This approach is intended to empower students in environmental engineering, chemical engineering, biotechnology, and related fields to not only understand the theoretical underpinnings but also apply them in practical, real-world scenarios.

I am deeply grateful to the students of my *Biological Processes in Environmental Engineering* course, whose feedback on early drafts sharpened this work. Their insights have been invaluable. As the sole author of this work, I have drawn on my own research and teaching to compile a resource that I hope will be both informative and inspiring. If you have ideas for improving this book or want to share your experiences, please contact me at varenyam@gmail.com. This book is as much yours as it is mine, and I look forward to hearing how it supports your journey.

Thank you for choosing *Notes on Biological Processes in Environmental Engineering*. I hope it serves as an indispensable resource, inspiring you as you explore and help shape sustainable solutions for our environment.

Shantou, China Varenyam Achal

Competing Interests The author has no competing interests to declare that are relevant to the content of this manuscript.

Contents

Chapter 1
The Basic Microbiology: An Engineer's Perspective

Learning Objectives:

1. *Understand prokaryotes, bacteria and importance of bacteria in environmental engineering.*
2. *Gain perspective on categorization of bacteria.*

The fundamental aspects of biological processes in environmental engineering, particularly in water and wastewater treatment, highlight the pivotal role of microorganisms in transforming and stabilizing pollutants. These biological processes are intricately tied to the activities of diverse microbial communities, making a foundational understanding of microbiology essential for engineers and scientists in this field.

1.1 Microbiology: The Backbone of Environmental Engineering

Microbiology, the scientific study of microorganisms, provides critical insights into the myriad of life forms invisible to the naked eye. Microorganisms, or microbes, include a vast array of organisms such as bacteria, archaea, viruses, fungi, prions, protozoa, and algae. Each plays a unique role in Earth's ecosystems, including engineered systems like wastewater treatment plants.

1.1.1 Bacteria and Archaea

Bacteria and archaea represent two of the three main cell lineages (Fig. 1.1), characterized by their simple, unicellular structure yet complex metabolic capabilities

V. Achal, *Notes on Biological Processes in Environmental Engineering*,
https://doi.org/10.1007/978-981-95-1610-0_1

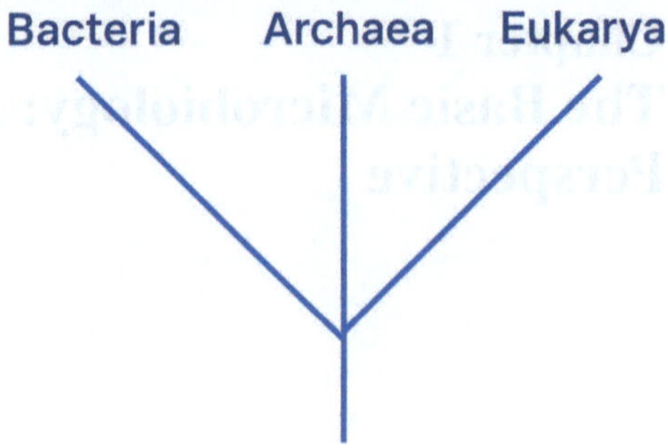

Fig. 1.1 The three cell lineages within microbes evolved from a common ancestor

(Reece et al. 2011). Bacteria are ubiquitous, thriving in every conceivable habitat on Earth. They play essential roles in decomposition, nitrogen fixation, and as pathogens or beneficial symbionts in human and animal guts.

Archaea, once thought to exist solely in extreme environments, are now known to inhabit diverse habitats and significantly contribute to global biogeochemical cycles, including methane production and consumption. Their unique biochemical properties make them critical players in environmental processes.

1.1.2 Viruses, Fungi, Prions, Protozoa, and Algae

Viruses, though not universally recognized as living organisms, influence microbial population dynamics and gene flow through mechanisms like transduction. Their roles in environmental engineering are significant, particularly in understanding microbial ecosystem dynamics.

Fungi contribute to the decomposition of organic matter and nutrient cycling. Due to their extensive enzyme systems, they are also instrumental in bioremediation efforts.

Prions, infectious proteins responsible for neurodegenerative diseases, present challenges in disease management and environmental contamination due to their resilience.

Protozoa serve as nutrient cyclers and predators of bacteria, influencing microbial community structure. Meanwhile, algae are pivotal for photosynthesis and oxygen production and are employed in wastewater treatment for nutrient removal and biomass generation.

1.1.3 Evolutionary Perspective

Understanding that all these diverse microbial groups evolved from a common ancestor provides an evolutionary lens to their functional diversity and adaptability.

This perspective enhances the appreciation of complex interactions and evolutionary pressures shaping microbial communities in both natural and engineered environments. By examining the evolutionary trajectories of these microorganisms, we can better understand how specific traits, such as metabolic versatility or resistance to environmental stressors, have emerged and been maintained. This knowledge is particularly valuable in engineering contexts, where optimizing microbial performance often involves leveraging their evolutionary adaptations. Additionally, an evolutionary perspective helps predict microbial responses to environmental changes, aiding in the design of robust and adaptable biological treatment systems.

1.2 Importance in Environmental Engineering

In water and wastewater treatment, microbiology knowledge is applied to harness the natural processes carried out by microorganisms. Biological treatment processes exploit the abilities of microbes to degrade organic pollutants, transform nutrients, and detoxify harmful substances, converting wastewater into effluent that can be safely discharged or reused.

Biodegradation: Specific bacteria and fungi excel at breaking down complex organic pollutants into simpler, non-toxic compounds, enabling effective pollutant removal.

Nutrient Removal: Microbial processes such as nitrification and denitrification, driven by specialized bacteria, are indispensable for removing excess nitrogen and phosphorus from wastewater. These processes help prevent eutrophication in natural water bodies.

Pathogen Reduction: Through mechanisms like competitive exclusion and natural die-off, microbial communities play a key role in reducing pathogenic organisms in treated water, enhancing safety and compliance with health standards.

1.3 Microbial Classification: Prokaryotes and Eukaryotes

Microbes are broadly categorized into two domains based on their cellular structures: prokaryotes and eukaryotes.

- **Prokaryotes (No Nucleus)**: The term "prokaryote" is derived from the Greek words πρό (pro, "before") and κάρυον (karyon, "nut" or "kernel"). Prokaryotes are single-celled organisms and represent the earliest and most primitive forms of life on Earth. These organisms lack a nucleus and other membrane-bound organelles within their cytoplasm. Prokaryotes are further classified into two distinct groups: bacteria and archaea.

Table 1.1 Table showing differences between prokaryotes and eukaryotes

Prokaryotic cells	Shared features	Eukaryotic cells
Simple and small	Cell membrane	Complex and large
0.1–5.0 μm in size	Cytoplasm	10–100 μm in size
No nucleus	Ribosomes	Nucleus present
Circular DNA	DNA as genetic material	Linear DNA
No organelles		Membrane-bound organelles
Bacteria, Archaea		Plants, Animals, Fungi, etc

- **Eukaryotes (True Nucleus)**: The term "eukaryote" originates from the Greek words εὖ (eu, "well" or "good") and κάρυον (karyon, "nut" or "kernel"). Eukaryotes may be unicellular or multicellular organisms. Unlike prokaryotes, eukaryotes possess nuclei and membrane-bound organelles. This category includes animals, plants, fungi, protists, and most algae.

The differences and similarities between prokaryotes and eukaryotes are summarized in Table 1.1.

1.4 Bacteria: Ubiquitous Prokaryotes

1.4.1 Morphology and Distribution

Bacteria (singular, bacterium) are among the most commonly encountered prokaryotes and are found in diverse environments such as soil, water, and as symbionts or pathogens within larger organisms. Typically, bacteria vary in size, ranging from 1 to 5 μm in length and 0.5–2 μm in width. A distinguishing feature of bacteria is their morphology, which varies by species and plays a role in their identification and function.

1.4.2 Nomenclature of Bacteria

The nomenclature of bacteria follows a binomial system introduced by Carl Linnaeus (1707–1778). In this system, a bacterial species name comprises a genus (generic name) and a specific epithet. The genus name, usually a Latin noun, identifies the broader classification, while the specific epithet, an adjective or noun, often describes a characteristic or origin of the species. The name can be derived from a person's name (frequently a microbiologist), source of isolation, a geographical location, a phenotypic characteristic (growth condition, colour, biochemical characteristics) or any other origin (Parte et al. 2020).

For example:

- ***Escherichia coli***: Named after Theodor Escherich, the species epithet "coli" refers to the colon, where it was first isolated.
- ***Acinetobacter refrigeratorensis***: Refers to its isolation from a refrigerator (Feng et al. 2014).
- ***Marinomonas shanghaiensis***: Denotes its geographical origin in Shanghai (Wei et al. 2019).

The genus name is capitalized, the specific epithet is in lowercase, and the entire name is italicized (e.g., *Bacillus cereus*). If italics are unavailable, the name is underlined. Official naming conventions are governed by the International Committee on Systematics of Prokaryotes (ICSP) and the International Code of Nomenclature of Bacteria.

1.4.3 Shapes of Bacteria

Bacteria exhibit four primary shapes, which are key to their classification and identification (Fig. 1.2):

- **Spherical (Cocci)**: Singular, coccus.
- **Rod-shaped (Bacilli)**: Singular, bacillus.
- **Comma-shaped (Vibrio)**.
- **Spiral-shaped (Spirochetes)**: Singular, spirochete.

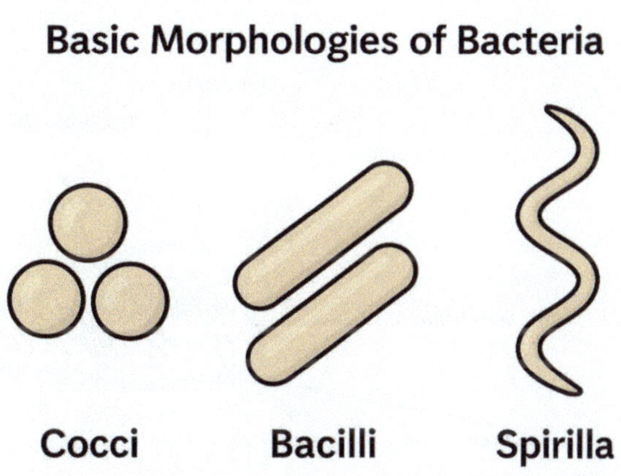

Fig. 1.2 An illustration showcasing the basic morphologies of bacteria, including cocci (spherical), bacilli (rod-shaped), and spirilla (spiral) shapes

1.5 Structure of Bacteria

The structure of bacteria (Fig. 1.3) is a remarkable adaptation that enables them to survive and thrive in diverse environments. Understanding bacterial structure is essential for comprehending their physiological roles, interactions, and applications in engineering and environmental sciences.

1.5.1 The Cell Envelope

The outermost layer of a bacterial cell, known as the cell envelope, is a multi-layered structure that provides shape, protection, and selective interaction with the external environment. It consists of the cell wall and the cytoplasmic membrane, with an additional outer membrane in Gram-negative bacteria.

- Cell Wall
  - The bacterial cell wall is a rigid, protective structure that maintains the cell's shape and prevents it from bursting due to osmotic pressure.
  - It is primarily composed of peptidoglycan, a complex polymer of sugars and amino acids. Peptidoglycan contains:

    N-acetylglucosamine (NAG)
    N-acetylmuramic acid (NAM)
    Short chains of amino acids, such as alanine, glutamic acid, and lysine or diaminopimelic acid.

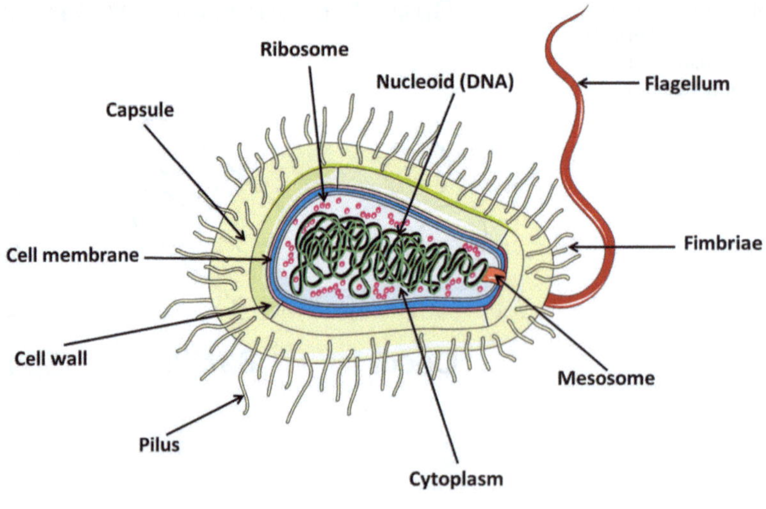

Fig. 1.3 Structure of a bacterial cell (from Jiménez-Jiménez et al. 2022)

- In Gram-positive bacteria, teichoic acids are present, contributing to cell wall rigidity and adhesion.
- In Gram-negative bacteria, an additional outer membrane is present, made of lipopolysaccharides (LPS) and phospholipids, which enhance structural integrity and protect against antibiotics.

- Cytoplasmic Membrane
  - Beneath the cell wall lies the cytoplasmic membrane, a lipid bilayer that acts as a permeability barrier. It regulates the transport of nutrients, waste products, and ions, maintaining cellular homeostasis.

1.5.2 Cytoplasm

The cytoplasm is a viscous, aqueous medium that houses various cellular components critical for bacterial metabolism and survival:

- Ribosomes: Protein synthesis centers.
- Mesosomes: Invaginations of the plasma membrane, involved in DNA replication and cell division.
- Inclusions: Storage structures for nutrients like polysaccharides, lipids, and phosphate granules (volutin).
- Vacuoles: Fluid-filled compartments that may regulate buoyancy or store nutrients.

1.5.3 Genetic Material

Bacteria lack a nucleus. Instead, their genetic material is organized as:

- Nucleoid: A region containing bacterial chromosome (DNA), which is typically circular and supercoiled.
- Plasmids: Extra-chromosomal DNA elements, often carrying genes for antibiotic resistance or metabolic capabilities.

1.5.4 External Features

- Capsule
  - Many bacteria secrete a capsule or slime layer around their surface.
  - Composed of polysaccharides, it enhances bacterial adhesion, biofilm formation, and protection against desiccation and immune responses.

- Flagella
  - Flagella are long, whip-like structures responsible for bacterial motility. They rotate like a propeller, enabling movement toward favorable environments (chemotaxis).
- Pili and Fimbriae
  - Pili: Longer, less numerous structures that facilitate genetic exchange through conjugation.
  - Fimbriae: Short, hair-like structures that promote adhesion to surfaces, playing a critical role in infection and biofilm formation.
- Spores
  - Some bacteria, such as *Bacillus* and *Clostridium*, form endospores under unfavorable conditions such as nutrient depletion or extreme temperatures.
  - Endospores contain calcium-dipicolinic acid, which makes them highly resistant to heat, desiccation, and chemicals, enabling survival in harsh environments.

1.6 Growth and Multiplication of Bacteria

Bacteria reproduce primarily by **binary fission**, a process in which a single cell divides into two identical daughter cells.

1.6.1 Generation Time

- The time required for a bacterial cell to divide and produce two daughter cells is called the **generation time** or **population doubling time**.
- Under optimal conditions, bacterial populations can grow exponentially, with a single cell potentially giving rise to over 10^{21} progeny in 24 h.

1.6.2 Growth Curve

When bacteria are cultured in a liquid medium, their growth follows a predictable pattern, represented by a growth curve (Fig. 1.4):

- **Lag Phase**
  - Bacteria acclimatize to the new environment. Metabolic activity is high, but cell division is minimal.

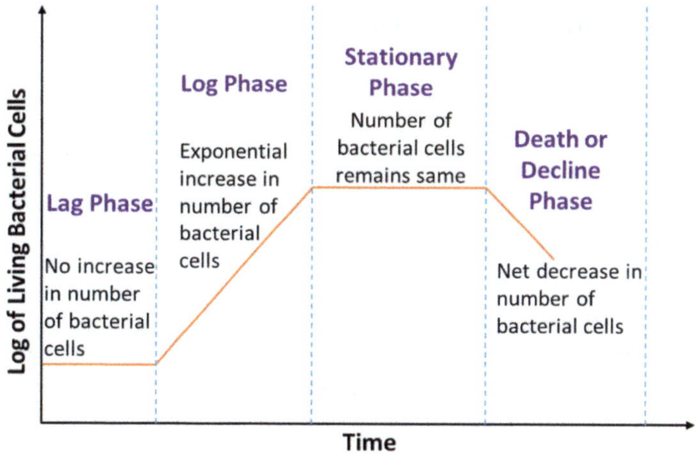

Fig. 1.4 Bacterial growth curve

- **Logarithmic (Exponential) Phase**
  - Cells divide rapidly, leading to exponential population growth. This is the phase of maximum metabolic activity.
- **Stationary Phase**
  - Growth slows as nutrients are depleted, and waste products accumulate. The number of new cells equals the number of dying cells.
- **Death Phase**
  - Nutrient exhaustion and toxic waste buildup lead to a decline in viable cell numbers.

1.6.3 Measuring Bacterial Growth

- **Direct Counting**
  - Counting cells using a microscope or devices like a Coulter counter.
- **Viable Counting**
  - Estimating live cells capable of forming colonies on solid media (colony-forming units, cfu/mL).
- **Turbidity Measurement**
  - Assessing optical density (OD) at 600 nm in a spectrophotometer, correlating with cell concentration.

1.7 Categorization of Bacteria

Bacteria can be classified nutritionally based on their energy and carbon sources, as illustrated in Fig. 1.5.

1.7.1 By Energy Source

- **Phototrophs**: Bacteria that derive their energy from sunlight.
- **Chemotrophs**: Bacteria that obtain energy from chemical compounds. If the source is organic, the bacteria are called **chemoorganotrophs**; if inorganic, they are termed **chemolithotrophs** (from the Greek *lithos* meaning "rocks").

1.7.2 By Carbon Source

- **Autotrophs**: Bacteria that utilize atmospheric carbon dioxide (inorganic carbon) as their carbon source.
- **Heterotrophs**: Bacteria that derive carbon from organic compounds.

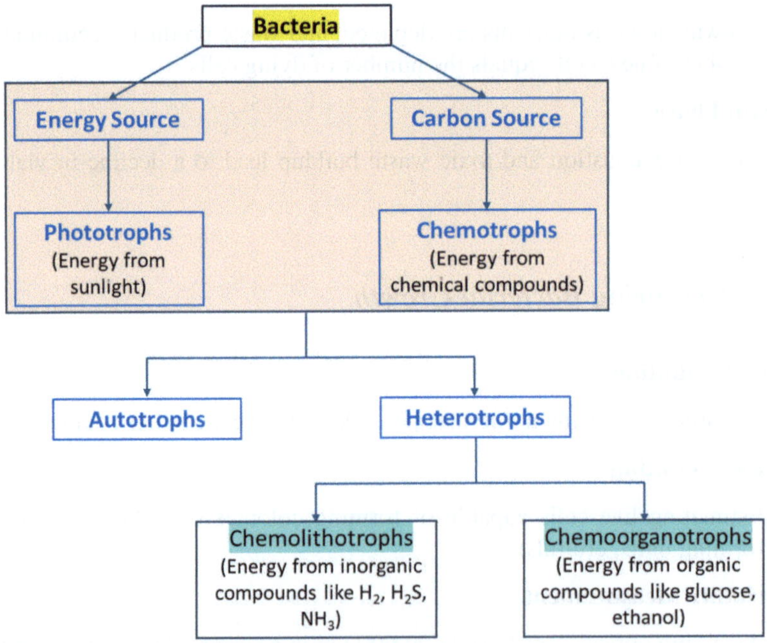

Fig. 1.5 Classification of bacteria based on energy and carbon sources

1.7.3 Combined Classifications

- **Chemolithotrophic bacteria** are typically autotrophic and are often referred to as **lithoautotrophs**.
- **Chemoorganotrophic bacteria** are commonly heterotrophic and are referred to as **heterotrophs**.
- **Phototrophs** are generally autotrophic and are often referred to as **photoautotrophs**.

Bacteria are further categorized based on the environmental conditions necessary for their growth, including oxygen requirements, temperature, and pH tolerance.

1.7.4 Oxygen Requirements

- **Aerobic Bacteria**: Require oxygen for survival and growth.
  - **Obligate aerobes** grow exclusively in the presence of oxygen.
  - **Facultative anaerobes** prefer oxygen but can grow without it.
- **Anaerobic Bacteria**: Grow in the absence of oxygen.
  - **Obligate anaerobes** are killed by oxygen.
  - **Aerotolerant anaerobes** or **microaerophilic bacteria** can survive in low oxygen concentrations. For example, genera like *Vibrio*, *Campylobacter*, *Neisseria*, and *Helicobacter* are microaerophilic (Perez-Perez and Blaser 1996).

Oxygen for microaerophiles is analogous to trace minerals like iron for humans: essential in small quantities but toxic in excess.

1.7.5 Temperature Preferences

Bacteria can also be classified by their optimal growth temperature, as shown in Fig. 1.6:

- **Psychrophilic Bacteria**: Thrive below 20 °C. Example: *Psychrobacter nivimaris* (Heuchert et al. 2004).
- **Mesophilic Bacteria**: Grow best at 25–40 °C. Example: *Escherichia coli.*
- **Thermophilic Bacteria**: Prefer temperatures of 55–80 °C. Example: *Bacillus stearothermophilus.*
- **Hyperthermophilic Bacteria**: Survive at temperatures above 80 °C. Example: *Methanopyrus kandleri* can grow at 122 °C (Takai et al. 2008).

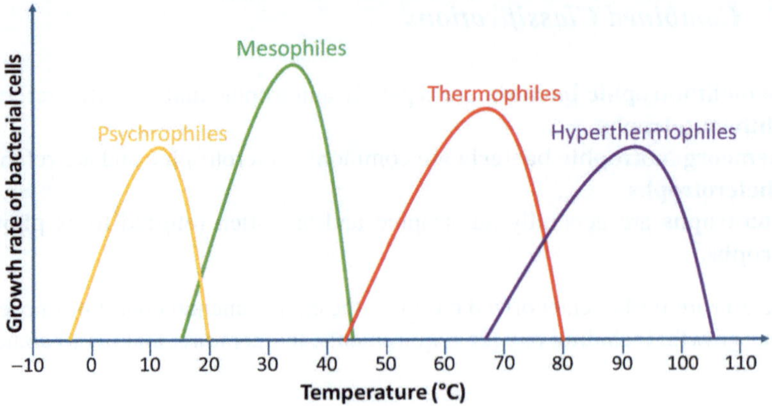

Fig. 1.6 Classification of bacteria based on their growth temperature

1.7.6 pH Preferences

Bacteria exhibit varying growth preferences based on pH, as illustrated in Fig. 1.7:

- **Acidophilic Bacteria**: Grow best under highly acidic conditions (pH $\leq$ 5.0). Example: *Acetobacter aceti*.
- **Neutrophilic Bacteria**: Prefer neutral pH environments (6.5–7.5). Example: *Escherichia coli*.
- **Alkaliphilic Bacteria**: Thrive at alkaline pH values ($\geq$9.0). Example: *Bacillus subtilis*.

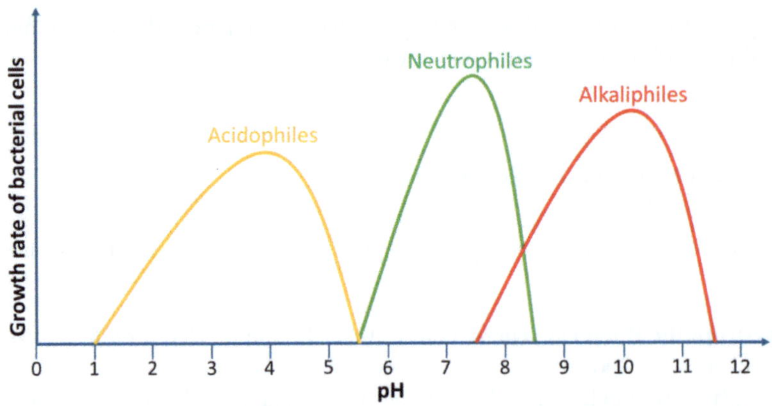

Fig. 1.7 Classification of bacteria based on their growth pH

1.7.7 *Optimal Growth Conditions*

Bacterial growth is most efficient at their **optimum growth temperature** and **pH**. Growth slows or ceases below the minimum or above the maximum thresholds. The word '*phile*' in terms such as "acidophile" or "alkaliphile" originates from the ancient Greek word *phileein*, meaning "to love."

This chapter provided an overview of the structural and functional characteristics of bacteria, starting with the composition of the bacterial cell envelope, including the rigid cell wall and cytoplasmic membrane, which protect the cell and maintain its shape. The cytoplasm, rich in ribosomes, mesosomes, inclusions, and vacuoles, was discussed alongside unique bacterial structures such as plasmids, capsules, flagella, pili, and fimbriae, as well as the formation of endospores for survival under adverse conditions.

The mechanisms of bacterial growth and reproduction were explored, emphasizing binary fission and the growth phases—lag, exponential, stationary, and death—along with methods to measure bacterial growth through total and viable counts.

The chapter also categorized bacteria based on their nutritional requirements, dividing them by energy sources (phototrophs and chemotrophs) and carbon sources (autotrophs and heterotrophs). Further classifications included environmental adaptations to oxygen availability, temperature, and pH. Aerobic and anaerobic bacteria were described along with psychrophilic, mesophilic, thermophilic, and hyperthermophilic species based on temperature preferences, and acidophilic, neutrophilic, and alkaliphilic bacteria based on pH preferences.

Further Readings

Feng G, Yang S, Wang Y, Yao Q, Zhu H (2014) *Acinetobacter refrigeratorensis* sp. nov., Isolated from a Domestic Refrigerator. Curr Microbiol 69:888–893

Heuchert A, Glöckner FO, Amann R, Fischer U (2004) *Psychrobacter nivimaris* sp. nov., a heterotrophic bacterium attached to organic particles isolated from the South Atlantic (Antarctica). Syst Appl Microbiol 27(4):399–406

Jiménez-Jiménez C, Moreno VM, Vallet-Regí M (2022) Bacteria-assisted transport of nanomaterials to improve drug delivery in cancer therapy. Nanomaterials 12(2):288

Parte AC, Sardà Carbasse J, Meier-Kolthoff JP, Reimer LC, Göker M (2020) List of Prokaryotic names with Standing in Nomenclature (LPSN) moves to the DSMZ. Int J Syst Evol Microbiol 70:5607–5612

Perez-Perez GI, Blaser MJ (1996) Campylobacter and helicobacter. In: Baron S (ed) Medical microbiology, 4th edn. University of Texas Medical Branch at Galveston, Galveston (TX). Chapter 23. Available from: https://www.ncbi.nlm.nih.gov/books/NBK8417/

Reece JB, Urry LA, Cain ML, Wasserman SA, Minorsky PV, Jackson RB (2011) Campbell biology, 9th edn. Pearson Education, Inc.

Takai K, Nakamura K, Toki T, Tsunogai U, Miyazaki J, Hirayama H, Nakagawa S, Nunoura T, Horikoshi K (2008) Cell proliferation at 122 °C and isotopically heavy CH_4 production by a hyperthermophilic methanogen under high-pressure cultivation. Proc Natl Acad Sci USA 105(31):10949–10954

Wei Y, Cao J, Mao H, Pei J, Liu R, Fang J (2019) *Marinomonas shanghaiensis* sp. nov., isolated from the junction between an ocean and a freshwater lake. Int J Syst Evol Microbiol 69(3):805–810

Chapter 2
Stoichiometry and Bacterial Energetics

Learning Objectives:

1. *Understand how to construct balanced mass equations—including elements, electrons, charge, and energy—for biological processes in wastewater treatment.*
2. *Apply stoichiometric principles and bacterial energetics to determine the chemical and nutrient requirements for bacterial growth and sludge production.*

One of the most important concepts in biological processes in environmental engineering including wastewater treatment is the mass balance. A mass balance is used to determine the amount of chemicals that must be supplied to satisfy the energy, nutrient, and environmental need of the bacteria to generate end products. In such mass balance equations, chemicals are electron acceptors (for example, oxygen), nutrients are nitrogen and phosphorus for biomass growth, and sulfuric acid or sodium hydroxide to maintain the pH of environment. Biomass here refers to the mass, weight or total quantity of bacteria. End products are excess bacteria in the form of sludge.

Balanced mass equations are based upon the concept of stoichiometry, as taught in chemistry. For biological processes such equations involve (i) oxidation and reduction of more than one species, (ii) bacteria act as catalysts and products of the reaction, and (iii) bacteria use the energy released from the reaction for cell synthesis and maintenance (bacterial energetics). For this reason, it is important to know how to write stoichiometric equations that balance elements, electrons, charge, and energy.

2.1 Stoichiometric Equation in Biological Processes

Porges et al. (1956) derived the **empirical formula for bacterial cells as $C_5H_7O_2N$**, which was one of the first to be used in the balancing of biological reactions. This is the most widely used formula as well. Even though bacterial cells contain a variety

V. Achal, *Notes on Biological Processes in Environmental Engineering*,
https://doi.org/10.1007/978-981-95-1610-0_2

of carbohydrates, proteins, fats, nucleic acids, and many elements other than carbon, hydrogen, oxygen, and nitrogen, Porges et al. (1956) only represented those four major elements because they are generally satisfactory for most practical purposes. A typical composition of bacterial cells is provided in Fig. 2.1.

Biological oxidation of wastewater containing casein was one of the first balanced biological reactions, presented by Porges et al. (1956).

$$C_8H_{12}O_3N_2 + 3O_2 \rightarrow C_5H_7O_2N + NH_3 + 3CO_2 + H_2O$$

As per this equation, for each 184 g of casein consumed by bacteria, 96 g of oxygen must be supplied for the reaction to proceed properly. The reaction produced 113 g of new bacterial cells, 17 g of ammonia (or 14 g of ammonia-N), 132 g of

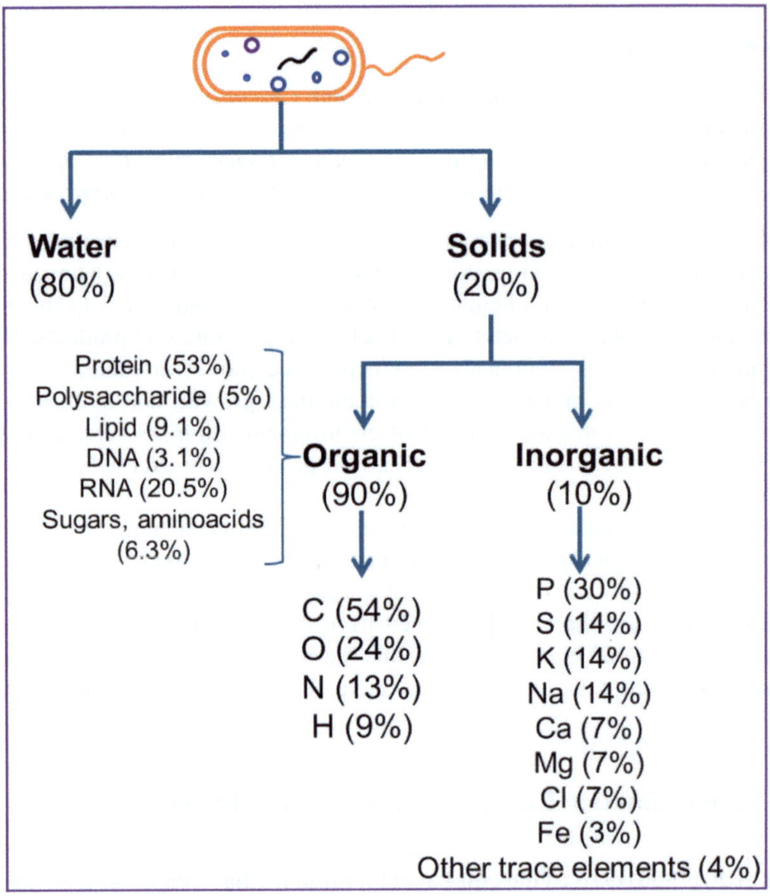

Fig. 2.1 Typical composition of a bacterial cell. (2003 Adapted from Madigan et al. (1997), Metcalf and Eddy (2003)

carbon dioxide, and 18 g of water. Based on this information, an engineer can design a biological treatment system for casein-containing wastewater.

Practice: How much oxygen should be supplied to treat casein-containing wastewater receiving influent at the rate of 500 kg/d? Also, determine the amount of biomass solids (dry sludge) that must be disposed.

Solution: From Eq. 2.1, to treat 184 kg/d casein, 96 kg/d oxygen is required that generated 113 kg/d of biomass. Thus, to treat 500 kg/d casein, 261 kg/d of oxygen must be supplied. The reaction will generate 307 kg/d of biomass that must be disposed.

2.2 Significance of the Empirical Formula of Bacterial Cell

- **Simplification of Complex Systems**: By boiling down the myriad of components within a bacterial cell to four key elements—carbon (C), hydrogen (H), oxygen (O), and nitrogen (N)—this formula allows for a more manageable approach to modeling and understanding microbial processes. It captures the essence of biomass composition, which is crucial for the analysis of growth, decay, and metabolic activities in microbial ecology and biotechnology applications.
- **Foundation for Stoichiometric Calculations**: The $C_5H_7O_2N$ formula serves as a basis for quantifying the conversion of substrates into biomass and by-products in microbial reactions. It enables the calculation of growth yields, substrate utilization rates, and the stoichiometry of nutrient requirements and product formation.
- **Versatility in Applications**: Despite its simplicity, the formula is widely applicable across various fields, including wastewater treatment, soil microbiology, and industrial bioprocessing. It helps in designing and optimizing processes like biodegradation, bioremediation, and the production of biochemicals, where microbial metabolism plays a central role.
- ***Application in Balancing Biological Reactions***: In the context of microbial metabolism, the stoichiometric equation involving $C_5H_7O_2N$ allows for the prediction of oxygen demand, nutrient requirements, and waste production. For instance, the oxidation of glucose by bacteria can be balanced using this formula to estimate the amount of oxygen required and the quantity of biomass produced, along with carbon dioxide and water as by-products.
- ***Limitations and Considerations***: While the empirical formula $C_5H_7O_2N$ is a powerful tool, it is essential to recognize its limitations. It does not account for the diversity of microbial species and their unique biochemical pathways. The actual composition of bacterial cells can vary depending on the species, growth conditions, and substrate availability. Therefore, while the formula provides a good

approximation for many practical purposes, specific applications may require adjustments or more detailed models to accurately reflect the biological process at hand.

Although $C_5H_7O_2N$ is the most general empirical formula used for bacterial cells, there are other empirical formula as well depending on substrate (food) used by bacteria for energy and cell synthesis in addition to other nutrients and environmental condition required for bacterial growth. From the empirical formula, one can calculate the oxygen required for full oxidation of the cellular carbon per unit weight of cells, termed as calculated oxygen demand (COD'). It is typically equal to the chemical oxygen demand (COD).

The empirical formula $C_nH_aO_bN_c$ was proposed for the average composition of bacterial cells, where:

$$n = \%C/12T, a = \%H/12T, b = \%O/16T, n = \%N/14T$$

and

$$T = \%C/12 + \%H + \%C/12 + \%O/16 + \%N/14$$

The calculated cell formula is generally normalized to c = 1 for N.

The COD′ for cell oxidation can be calculated by following formula:

$$C_nH_aO_bN_C + \left(\frac{2n + 0.5a - 1.5c - b}{2}\right)O_2 \rightarrow nCO_2 + cNH_3 + \left(\frac{a - 3c}{2}\right)H_2O$$

$$COD'/Weight = \frac{(2n + 0.5a - 1.5c - b)16}{12n + a + 16b + 14c}$$

Practice: What would be the empirical formula and.

COD′/Weight for the bacterial cell containing 43.5% C, 9.8% H, 27.4% O, and 7.5% N?

Solution:

$$T = 43.5/12 + 9.8 + 27.4/16 + 7.5/14 = 15.68$$
$$n = 43.5/(12 \times 15.68) = 0.23; a = 9.8/15.68 = 0.63$$
$$b = 27.4/(16 \times 15.68) = 0.11; c = 7.5/(14 \times 15.68) = 0.034$$

In order to normalize to c = 1 for N, divide by 0.034, the following empirical formula for the cells is obtained: $C_{6.8}H_{18.5}O_{3.2}N$.

The COD' to organic weight ratio is then equal to $(2 \times 6.8 + 0.5 \times 18.5 - 1.5 - 0.11) \times 16/(12 \times 6.8 + 18.5 + 16 \times 3.2 + 14) =$ 2.06 g COD'/g cells.

It means 2.06 g of oxygen is needed for the complete oxidation of 1 g of cells.

2.3 Substrate Partitioning and Cellular Yield

Substrates in wastewater treatment using biological processes are often pollutants that are required to remove. It acts as nutrition for bacteria to provide energy and for cell synthesis. Substrate is also an electron donor and when bacteria use it, a portion of its electrons (f_e^0) is transferred to the electron acceptor to provide energy, while other portion of electrons (f_s^0) into cell synthesis/bacterial cells (Fig. 2.2). As electron flows generate the cell's energy, this partitioning is often expressed as electron equivalents (e^-eq).

The fraction can be f_s^0 converted into mass units, termed true yield (Y).

$$Y = \frac{f_s^0 M_C}{8n_e}$$

where,

Y	g cell produced per g COD.
f_s^0	g cell produced per g COD′.
M_c	empirical formula weight of cells.
n_e	number of electron eq. in a mole of cells

$$Y \propto f_s^0$$

In the above equation, 8 is e^-eq. As in redox relationship, one oxygen transfer is equivalent to 2 e^- transfer, e^-eq is equivalent to 8 g of COD.

The net yield (will be discussed in Chap. 3) is always less than Y, because some of the electrons originally present in the substrate must be consumed for energy of

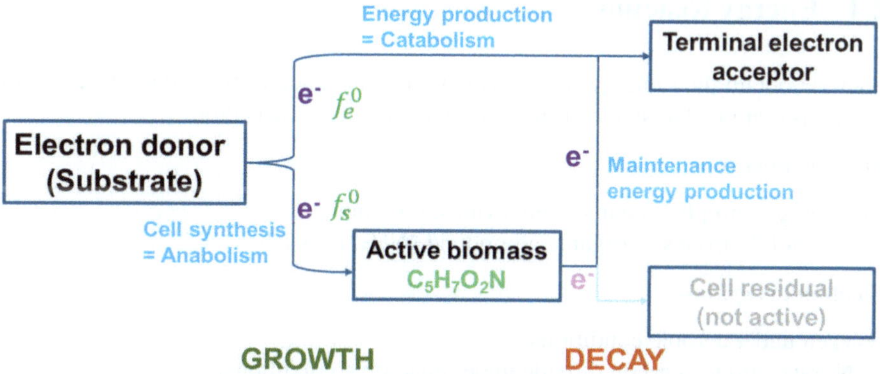

Fig. 2.2 Utilization of electron donor (substrate) by bacteria for energy production and cell synthesis [f_e^0 = portion of electrons of an e^- donor substrate to provide energy, f_s^0 = portion of electrons of an e^- donor substrate for cell synthesis]

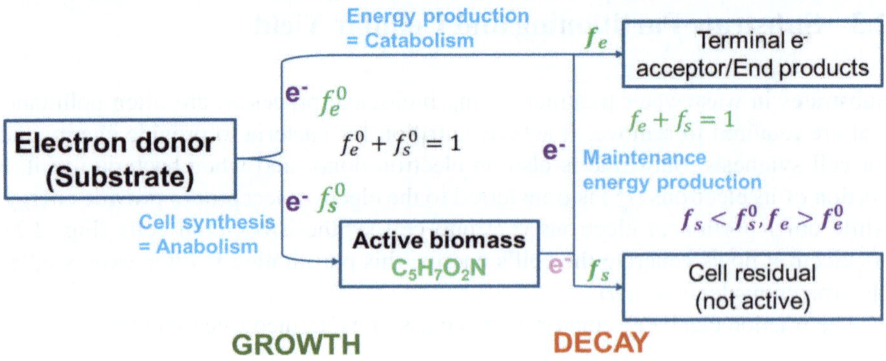

Fig. 2.3 Utilization of electron donor (substrate) by bacteria for energy production and cell synthesis, finally leading to end products formation and cell decay

maintenance. It makes the portion of electrons used for synthesis as f_s rather than f_s^0, and the portion of electrons for energy production as f_e rather than f_e^0. Accordingly, utilization of substrate by bacteria is finally presented as in Fig. 2.3.

In short, the observed fraction of electrons for cell synthesis (f_s^0) and the theoretical fraction of electrons for cell synthesis (f_s) are related through the concept of biomass yield. The observed f_s^0 represents the actual fraction of electrons that are used for building cell biomass, while fs represents the maximum theoretical fraction that could be used based on the stoichiometry of the process. Factors like energy lost to maintenance and other metabolic pathways can lead to f_s^0 being lower than f_s. On other hand, the f_e provides a theoretical estimate of electron utilization for energy production, the observed fraction (f_e^0) reflects the actual efficiency under specific experimental conditions.

2.4 Energy Reactions

Bacteria obtain their energy for growth and maintenance from oxidation–reduction reactions that involve an electron donor and an electron acceptor.

Electron donor–

Chemoorganotrophs: organic matter (most common).

Chemolithotrophs: inorganic compound (NH_3, H_2S).

Electron acceptor–

Oxygen under aerobic conditions.

Nitrate, sulfate, carbon dioxide under anaerobic conditions.

In fermentation reaction, organic matter is used as both the electron donor and acceptor.

Table 2.1 Stoichiometry of microbial glucose oxidation with various electron acceptors (ordered by acceptor preference)

Metabolic process	Balanced reaction	$\Delta G^{o'}$ (kJ/mol Glucose)
Aerobic oxidation	$C_6H_{12}O_6 + 6O_2 \rightarrow 6CO_2 + 6H_2O$	–2880
Denitrification	$5C_6H_{12}O_6 + 24NO_3^- + 24H^+ \rightarrow$ $30CO_2 + 42H_2O + 12N_2$	–2720
Sulfate reduction	$2C_6H_{12}O_6 + 6SO_4^- + 9H^+ \rightarrow$ $12CO_2 + 12H_2O + 3H_2S + 3HS^-$	–492
Methanogenesis	$C_6H_{12}O_6 \rightarrow 3CO_2 + 3CH_4$	–428
Ethanol fermentation	$C_6H_{12}O_6 \rightarrow 2CO_2 + 2C_2H_5OH$	–244

Preference for electron acceptors by bacteria:

oxygen > nitrate > sulfate > carbon dioxide > organics (fermentation)

The various energy reactions with glucose as electron donor and different electron acceptors are presented in Table 2.1.

Based on reactions in Table 2.1, aerobic organisms require to send few electrons from donor to acceptor (oxygen) to generate energy for biomass synthesis (f_e^0 small, f_s^0 large). Thus, aerobic bacteria will have higher true yield (Y) than anaerobic bacteria ($Y \propto f_s^0$). In other words, aerobic bacteria grow faster than anaerobic bacteria.

2.5 Constructing Energy Reactions

Writing energy reaction is the first step towards developing stoichiometric equation for bacterial growth. A reaction is made up of oxidation half-reaction, reduction half-reaction, and their sum. Tables 2.2 and 2.3 list half-reactions that are very important to write energy as well as biological reactions.

Practice: Construct denitrification reaction of glucose oxidation with nitrate, as presented in Table 2.1.

Solution:

Step 1—Write oxidation half-reaction for glucose

$$\frac{1}{24}C_6H_{12}O_6 + \frac{1}{4}H_2O \rightarrow \frac{1}{4}CO_2 + H^+ + e^-$$

Table 2.2 Balanced inorganic redox half-reaction equations at pH 7.0*

#	Reduced-oxidized form of compounds	Half-reaction	$\Delta G^{o\prime}$ (kJ/e⁻ eq)
1	Ammonium-Nitrate	$\frac{1}{8}NO_3^- + \frac{5}{4}H^+ + e^- = \frac{1}{8}NH_4^+ + \frac{3}{8}H_2O$	–35.11
2	Ammonium-Nitrite	$\frac{1}{6}NO_2^- + \frac{4}{3}H^+ + e^- = \frac{1}{6}NH_4^+ + \frac{1}{3}H_2O$	–32.93
3	Ammonium-Nitrogen	$\frac{1}{6}N_2 + \frac{4}{3}H^+ + e^- = \frac{1}{3}NH_4^+$	26.70
4	Ferrous-Ferric	$Fe^{3+} + e^- = Fe^{2+}$	–74.27
5	Hydrogen-H^+	$H^+ + e^- = \frac{1}{2}H_2$	39.87
6	Nitrite-Nitrate	$\frac{1}{2}NO_3^- + H^+ + e^- = \frac{1}{2}NO_2^- + \frac{1}{2}H_2O$	–41.65
7	Nitrogen-Nitrate	$\frac{1}{5}NO_3^- + \frac{6}{5}H^+ + e^- = \frac{1}{10}N_2 + \frac{3}{5}H_2O$	–72.20
8	Nitrogen-Nitrite	$\frac{1}{3}NO_2^- + \frac{4}{3}H^+ + e^- = \frac{1}{6}N_2 + \frac{2}{3}H_2O$	–92.56
9	Sulfide-Sulfate	$\frac{1}{8}SO_4^{2-} + \frac{19}{16}H^+ + e^- = \frac{1}{16}H_2S + \frac{1}{16}HS^- + \frac{1}{2}H_2O$	20.85
10	Sulfide-Sulfite	$\frac{1}{6}SO_3^{2-} + \frac{5}{4}H^+ + e^- = \frac{1}{12}H_2S + \frac{1}{12}HS^- + \frac{1}{2}H_2O$	11.03
11	Sulfite-Sulfate	$\frac{1}{2}SO_4^{2-} + H^+ + e^- = \frac{1}{2}SO_3^{2-} + \frac{1}{2}H_2O$	50.30
12	Sulfur-Sulfate	$\frac{1}{6}SO_4^{2-} + \frac{4}{3}H^+ + e^- = \frac{1}{6}S + \frac{2}{3}H_2O$	19.15
13	Thiosulfate-Sulfate	$\frac{1}{4}SO_4^{2-} + \frac{5}{4}H^+ + e^- = \frac{1}{8}S_2O_3^{2-} + \frac{5}{8}H_2O$	23.58
14	Water-Oxygen	$\frac{1}{4}O_2 + H^+ + e^- = \frac{1}{2}H_2O$	–78.72

*Thermodynamic data adapted from multiple sources including (Rittmann and McCarty 2011). Values compiled and reformatted for clarity

Step 2—Write reduction half-reaction for nitrate

$$\frac{1}{5}NO_3^- + \frac{6}{5}H^+ + e^- \rightarrow \frac{1}{10}N_2 + \frac{3}{5}H_2O$$

Above reactions are written for one electron equivalent.

Table 2.3 Balanced organic redox half-reaction equations at pH 7*

#	Compounds	Half-reaction	$\Delta G^{\circ'}$ (kJ/e⁻ eq)
1	Acetate	$\frac{1}{8}CO_2 + \frac{1}{8}HCO_3^- + H^+ + e^- = \frac{1}{8}CH_3COO^- + \frac{3}{8}H_2O$	27.4
2	Alanine	$\frac{1}{6}CO_2 + \frac{1}{12}HCO_3^- + \frac{1}{12}NH_4^+ + \frac{11}{12}H^+ + e^- = \frac{1}{12}CH_3CHNH_2COO^- + \frac{5}{12}H_2O$	31.37
3	Benzoate	$\frac{1}{5}CO_2 + \frac{1}{30}HCO_3^- + H^+ + e^- = \frac{1}{30}C_6H_5COO^- + \frac{13}{30}H_2O$	27.34
4	Citrate	$\frac{1}{6}CO_2 + \frac{1}{6}HCO_3^- + H^+ + e^- = \frac{1}{18}(COO^-)CH_2COH(COO^-)CH_2COO^- + \frac{4}{9}H_2O$	33.08
5	Ethanol	$\frac{1}{6}CO_2 + H^+ + e^- = \frac{1}{12}CH_3CH_2OH + \frac{1}{4}H_2O$	31.18
6	Formate	$\frac{1}{12}HCO_3^- + H^+ + e^- = \frac{1}{2}HCOO^- + \frac{1}{2}H_2O$	39.19
7	Glucose	$\frac{1}{4}CO_2 + H^+ + e^- = \frac{1}{24}C_6H_{12}O_6 + \frac{1}{4}H_2O$	41.35
8	Glutamate	$\frac{1}{6}CO_2 + \frac{1}{9}HCO_3^- + \frac{1}{18}NH_4^+ + H^+ + e^- = \frac{1}{18}COOHCCH_2CH_2CHNH_2COO^- + \frac{4}{9}H_2O$	30.93
9	Glycerol	$\frac{3}{14}CO_2 + H^+ + e^- = \frac{1}{14}CH_2OHCHOHCH_2OH + \frac{3}{14}H_2O$	38.88
10	Glycine	$\frac{1}{6}CO_2 + \frac{1}{6}HCO_3^- + \frac{1}{6}NH_4^+ + H^+ + e^- = \frac{1}{6}CH_2NH_3COOH + \frac{1}{2}H_2O$	39.8
11	Lactate	$\frac{1}{6}CO_2 + \frac{1}{12}H^+ + e^- = \frac{1}{12}CH_3CHOHCOO^- + \frac{1}{3}H_2O$	32.29
12	Methane	$\frac{1}{8}CO_2 + H^+ + e^- = \frac{1}{8}CH_4 + \frac{1}{4}H_2O$	23.53
13	Methanol	$\frac{1}{6}CO_2 + H^+ + e^- = \frac{1}{6}CH_3OH + \frac{1}{6}H_2O$	36.84
14	Palmitate	$\frac{15}{19}CO_2 + \frac{1}{92}HCO_3^- + H^+ + e^- = \frac{1}{92}CH_3(CH_2)_{14}COO^- + \frac{31}{92}H_2O$	27.26
15	Propionate	$\frac{1}{7}CO_2 + \frac{1}{14}HCO_3^- + H^+ + e^- = \frac{1}{14}CH_3CH_2COO^- + \frac{5}{14}H_2O$	27.63
16	Pyruvate	$\frac{1}{5}CO_2 + \frac{1}{10}HCO_3^- + H^+ + e^- = \frac{1}{10}CH_3COCOO^- + \frac{2}{5}H_2O$	35.09
17	Succinate	$\frac{1}{7}CO_2 + \frac{1}{7}HCO_3^- + H^+ + e^- = \frac{1}{14}(CH_2)_2(COO^-)_2 + \frac{3}{7}H_2O$	29.09

*Thermodynamic data adapted from multiple sources including (Rittmann and McCarty 2011). Values compiled and reformatted for clarity

One of aims of writing any such equation should be to cancel all electrons, means there must be no free electrons (e^-) in the overall reaction.

Step 3—Overall balanced reaction (sum of oxidation and reduction half-reactions)

$$\frac{1}{24}C_6H_{12}O_6 + \frac{1}{5}NO_3^- + \frac{1}{5}H^+ \rightarrow \frac{1}{4}CO_2 + \frac{7}{20}H_2O + \frac{1}{10}N_2$$

From above equations, by multiplying by 120 to both sides:

$$5C_6H_{12}O_6 + 24NO_3^- + 24H^+ \rightarrow 30CO_2 + 42H_2O + 12N_2$$

Organic half-reaction can further be customized with equation below.

$$\frac{(n-c)}{d}CO_2 + \frac{c}{d}HCO_3^- + \frac{c}{d}NH_4^+ + H^+ + e^- = \frac{1}{d}C_nH_aO_bN_C + \left(\frac{2n-b+c}{d}\right)H_2O$$

where, d = (4n + a − 2b − 3c)

Practice: Express cellular yield when bacterial cells $C_5H_7O_2N$ get synthesized with ammonia as the nitrogen source according to reaction below.

$$\frac{1}{5}CO_2 + \frac{1}{20}HCO_3^- + \frac{1}{20}NH_4^+ + H^+ + e^- = \frac{1}{20}C_5H_7O_2N + \frac{9}{20}H_2O$$

Solution: The reaction can be written as below for one mol of cells.

$$4CO_2 + HCO_3^- + NH_4^+ + H^+ + e^- = C_5H_7O_2N + 9H_2O$$

According to reaction, M_c = 113 g cells/mol cells, n_e = 20 e^-eq/mol cellse^-eq/mol cells

$$Y = \frac{f_s^0 M_C}{8n_e}$$

$$Y = \frac{f_s^0 \times 113}{8 \times 20} = 0.706 f_s^0 \text{ g cells/g COD}$$

2.6 Overall Reactions for Biological Growth

Bacterial growth involves two basic reactions: (i) for energy production and (ii) for cellular synthesis. Based on this electron patronizing and adding two half-reactions, overall balanced equation can be written as discussed above. However, for writing overall reaction for bacterial growth, or in short, biological reaction, half-reaction for cell synthesis (R_c) is also required. Table 2.4 lists major cell synthesis half-reactions. Ammonium is always the most preferred nitrogen source for bacteria, although bacteria can use other nitrogen source as listed in Table 4 in its absence.

The overall biological reaction is made up of energy reaction (R_e) involving donor half-reaction (R_d) and acceptor half-reaction (R_a) in addition to synthesis reaction (R_s).

Energy reaction, $R_e = R_a - R_d$

Synthesis reaction, $R_s = R_c - R_d$

Finally, to write the general equation for constructing stoichiometric equations for bacterial synthesis and growth, there is requirement to consider electron partitioning, f_e for energy reaction and f_s for synthesis reaction (McCarty 1975; Rittmann and McCarty 2001). It makes overall biological reaction (R) as:

$$R = f_e R_e + f_s R_s$$

$$= f_e(R_a - R_d) + f_s(R_c - R_d)$$

$$= f_e R_a + f_s R_c - R_d(f_e + f_s)$$

Since, $f_e + f_s = 1$

$$R = f_e R_a + f_s R_c - R_d$$

Table 2.4 Half-reactions of cell synthesis (R_c)

Reaction parameter	Half-reaction
When ammonium is nitrogen source	$\frac{1}{20}NH_4^+ + \frac{1}{5}CO_2 + \frac{1}{20}HCO_3^- + H^+ + e^- = \frac{1}{20}C_5H_7O_2N + \frac{9}{20}H_2O$
When nitrate is nitrogen source	$\frac{1}{28}NO_3^- + \frac{5}{28}CO_2 + \frac{29}{28}H^+ + e^- = \frac{1}{28}C_5H_7O_2N + \frac{11}{28}H_2O$
When nitrite is nitrogen source	$\frac{1}{26}NO_2^- + \frac{5}{26}CO_2 + \frac{27}{26}H^+ + e^- = \frac{1}{26}C_5H_7O_2N + \frac{10}{26}H_2O$
When dinitrogen is nitrogen source	$\frac{1}{46}N_2 + \frac{5}{23}CO_2 + H^+ + e^- = \frac{1}{23}C_5H_7O_2N + \frac{8}{23}H_2O$

R represents net consumption of reactants and production of products when the microorganisms consume one e-equivalent of electron donor (an electron equivalent basis). This equation is used to construct a wide variety of stoichiometric equations for bacterial synthesis and growth.

To write the overall biological reaction, f_eR_a, f_sR_c and $-R_d$ are added. There should not be confusion from negative sign placed before R_d, as it is because the donor is oxidized.

Practice: Formulate the overall biological reaction for the fermentation of glucose to ethanol, using ammonium as the nitrogen source for biomass synthesis and assuming an f_s equals 0.22.

Solution: We know, overall biological reaction,

$$R = f_eR_a + f_sR_c - R_d$$

From $f_e + f_s = 1, f_e = 1 - f_s = 1 - 0.22 = 0.78$. We need to identify electron acceptor and donor for this case. Here, electron donor is glucose that will be used by bacteria for ethanol fermentation. Since this is fermentation process, electron acceptor would be carbon dioxide. Half-cell reaction for cell synthesis (R_c) is also required that can be written with ammonium as nitrogen source.

First step—Write R_a

$$R_a : \frac{1}{6}CO_2 + H^+ + e^- \rightarrow \frac{1}{12}C_2H_5OH + \frac{1}{4}H_2O$$

Second step—Write R_cR_c

$$R_c : \frac{1}{5}CO_2 + \frac{1}{20}HCO_3^- + \frac{1}{20}NH_4^+ + H^+ + e^- \rightarrow \frac{1}{20}C_5H_7O_2N + \frac{9}{20}H_2O$$

Third step—Write $-R_d$

$$-R_d : \frac{1}{24}C_6H_{12}O_6 + \frac{1}{4}H_2O \rightarrow \frac{1}{4}CO_2 + H^+ + e^-$$

Now, we know:

$$R = f_eR_a + f_sR_c - R_d$$

Fourth step—Multiply R_a equation with f_e i.e., 0.78

$$f_eR_a : 0.13CO_2 + 0.78H^+ + 0.78e^- \rightarrow 0.065C_2H_5OH + 0.195H_2O$$

Fifth step—Multiply R_c equation with f_s i.e., 0.22

$$f_sR_c : 0.044CO_2 + 0.011HCO_3^- + 0.011NH_4^+$$
$$+0.22H^+ + 0.22e^- \rightarrow 0.011C_5H_7O_2N + 0.099H_2O^{\cdot}$$

Sixth step—Write $-R_d$ in decimal form.

$$-R_d : 0.042C_6H_{12}O_6 + 0.25H_2O \rightarrow 0.25CO_2 + H^+ + e^-$$

Seventh step—Add f_eR_a, f_sR_c and $-R_d$ equations.

$$f_eR_a : 0.13CO_2 + 0.78H^+ + 0.78e^- \rightarrow 0.065C_2H_5OH + 0.195H_2O$$
$$f_sR_c : 0.044CO_2 + 0.011HCO_3^- + 0.011NH_4^+ + 0.22H^+ + 0.22e^-$$
$$\rightarrow 0.011C_5H_7O_2N + 0.099H_2O$$
$$-R_d : 0.042C_6H_{12}O_6 + 0.25H_2O \rightarrow 0.25CO_2 + H^+ + e^-$$

$$R : 0.042C_6H_{12}O_6 + 0.011HCO_3^- + 0.011NH_4^+ \rightarrow 0.011C_5H_7O_2N$$
$$+0.065C_2H_5OH + 0.076CO_2 + 0.044H_2$$

2.7 Bacterial Energetics

Bacterial energetics involves the study of how bacteria obtain, convert, and use energy to sustain their life processes. These microscopic organisms have evolved diverse metabolic pathways to harness energy from various sources, including light, organic compounds, and inorganic chemicals. The energy extracted is primarily used for growth, reproduction, movement, and maintenance of cellular functions. Bacteria can perform both aerobic and anaerobic respiration, or even photosynthesis in the case of phototrophic bacteria, to meet their energy requirements. The efficiency of bacterial energetics not only supports their survival in a wide range of environments but also plays a crucial role in ecological balance, biogeochemical cycles, and industrial applications such as wastewater treatment and bioremediation. Understanding bacterial energetics is essential for microbiology, environmental science, and biotechnology, offering insights into the fundamental mechanisms of life and the potential for harnessing bacterial capabilities for human benefit.

Biodegradation, deterioration, oxidation–reduction, fermentation, putrefaction, and decay are typical biological processes associated with bacteria. Bacterial (or microbial growth), however, is dependent upon the input of energy and the growth yield can be used to estimate the efficiency of energy generation. Bacteria carry out oxidation–reduction reactions in order to obtain energy for growth and cell maintenance. The amount of energy released per electron equivalent of an electron donor oxidized varies considerably from reaction to reaction.

It is simple to relate electron equivalents to measurements commonly used in environmental engineering, including COD.

$$1\ \text{electron equivalent}\ (e^{-}eq) = 8\text{gO}_2$$

$$\frac{1}{4}\text{O}_2 + \text{H}^+ + \text{e}^- = \frac{1}{2}\text{H}_2\text{O}$$

Practice:

A wastewater stream contains 15 g/L of glycerol. Estimate:

- The electron-equivalents per liter ($e^{-}eq/L$)
- The corresponding chemical oxygen demand (COD, g/L)

Solution:

$$\frac{3}{14}\text{CO}_2 + \text{H}^+ + \text{e}^+ \rightarrow \frac{1}{14}\text{CH}_2\text{OHCHOHCH}_2\text{OH} + \frac{3}{14}\text{H}_2\text{O}$$

Step 1: Determine the Equivalent Weight of Glycerol:

- Ethanol Glycerol has a molecular weight of 92 g/mol.
- Based on half-reaction of glycerol, it is given that 1 mol of glycerol corresponds to 14 electron equivalents ($e^{-}eq$).

$$\text{Equivalent weight} = \frac{92\frac{g}{\text{mol}}}{14\frac{e^{-}eq}{\text{mol}}} = 6.57\frac{\text{g}}{e^{-}eq}$$

Step 2: Calculate the Electron Equivalents per Liter:

- The wastewater contains 15 g glycerol per liter.
- The number of electron equivalents in 15 g of glycerol is:

$$\frac{e^{-}eq}{L} = \frac{15\frac{\text{g}}{\text{L}}}{6.57\frac{\text{g}}{e^{-}eq}} = 2.28\frac{e^{-}eq}{\text{L}}$$

Step 3: Determine the COD′:

- It is known that each electron equivalent corresponds to an oxygen demand of 8 g COD per electron equivalent.
- Therefore, the COD′ is:

$$COD' = 8\,\text{gO}_2/e^{-}eq \times 2.28\,e^{-}eq/\text{L} = 18.24\,\text{g/L}$$

Answer:

The wastewater contains approximately 2.28 e^{-}eq/L, and the corresponding COD′ is 18.24 g/L.

2.8 Change in Gibbs Energy ($\Delta G°$)

The concept of change in Gibbs energy, denoted as $\Delta G°$, is pivotal in the field of thermodynamics and energetics, including bacterial energetics, serving as a fundamental measure to predict the spontaneity and equilibrium of chemical and physical processes. It embodies the maximum amount of non-expansion work that can be extracted from a system at constant temperature and pressure, making it an indispensable tool for chemists, physicists, and engineers alike.

Understanding Gibbs Energy

Gibbs energy, named after Josiah Willard Gibbs, integrates the enthalpy (H), temperature (T), and entropy (S) of a system into a single value, calculated using the equation:

$$\Delta G = \Delta H - T\Delta S$$

$\boldsymbol{\Delta H}$ is the enthalpy change, reflecting heat absorbed (endothermic) or released (exothermic).
$\boldsymbol{\Delta S}$ is the entropy change, indicating the degree of disorder created or removed.
$\boldsymbol{T}$ is the absolute temperature in kelvins.

Standard Gibbs Energy Change ($\boldsymbol{\Delta G°}$)

The notation $\Delta G°$ refers specifically to the Gibbs energy change under standard conditions, which typically mean 1 mol of substances, a pressure of 1 atmosphere, and a temperature of 25 °C (298 K). This standardization allows for the comparison of the energetics of reactions under consistent conditions. In biological contexts, the standard conditions often include a pH of 7, denoted as $\Delta G^{\circ'}$, to reflect the physiological conditions more accurately.

Under these conventions:

$\boldsymbol{\Delta G°} < 0 \Rightarrow$ products are thermodynamically favored under standard conditions.
$\boldsymbol{\Delta G°} > 0 \Rightarrow$ reactants are favored; external energy is needed to drive the reaction forward.
$\boldsymbol{\Delta G°} = 0 \Rightarrow$ the reaction mixture of standard-state species lies precisely at equilibrium.

Biological Context

For biological processes, often the standard Gibbs energy is given for pH 7, denoted by adding a prime (′) to the symbol for the Gibbs energy. In microbial and enzymatic systems, understanding $\Delta G^{o'}$ is crucial for mapping out metabolic pathways. Each step's $\Delta G^{o'}$ tells us whether the transformation releases usable energy (driving cellular work) or requires coupling to other exergonic reactions (e.g., ATP hydrolysis) to proceed. By examining both ΔH° and ΔS° terms, one gains insight into how temperature shifts or concentration changes can alter metabolic fluxes and equilibrium positions in living cells.

In order to understand energetics further, a calculation-based practice is given below.

Practice:

$$C_6H_{12}O_6(s) + 6O_2(g) \rightarrow 6H_2O(l) + 6CO_2(g)$$

a. Considering above reaction of glucose oxidation, how many mols of electrons were donated by 1 mol of glucose?

Solution a:

Write two half reactions to complete glucose oxidation.

$$\begin{array}{ll} C_6H_{12}O_6 + 6H_2O \rightarrow 6CO_2 + 24H^+ + 24e^- & \\ O_2 + 4H^+ + 4e^- \rightarrow 2H_2O & [\times 6] \\ \hline C_6H_{12}O_6 + 6O_2 \rightarrow 6H_2O + 6CO_2 & \end{array}$$

From balancing above equations, it is clear that there are 24 mols of electrons donated by 1 mol of glucose.

b. What would be the Gibbs free energy change of the reaction per mol electrons?

Solution b:

$$\Delta G_r^0 = \frac{-2878 \frac{\text{kJ}}{\text{mol Glucose}}}{24 \frac{\text{mol e}^-}{\text{mol Glucose}}} = -120 \frac{\text{kJ}}{\text{mol e}^-}$$

c. What would be the Gibbs free energy change of the reaction per one electron?

Solution c:

$$\Delta G_r = \frac{\Delta G_r^0}{NA} = -\frac{120}{6.02 \times 1023} = -2 \times 10^{-22} \frac{\text{kJ}}{\text{e}^-}$$

d. What would be the Gibbs free energy change of the reaction per electric charge?

$$\Delta G_r = \frac{\Delta G_r^0}{q_{e^-}} = \frac{-2 \times 10^{-22} \frac{\text{kJ}}{\text{e}^-}}{1.6 \times 10^{-19} \frac{\text{C}}{\text{e}^-}} = -0.00124 \frac{\text{kJ}}{\text{C}} = -1.24 \frac{\text{kJ}}{\text{C}} = -1.24\,\text{V}$$

2.9 Gibbs Energy Change During a Chemical Reaction

For the general gas-phase reaction:

$$\alpha A + \beta b \rightleftharpoons \gamma C + \delta D$$

we can make the connection between Gibbs free energy, the reaction quotient, and the equilibrium constant very explicit by invoking the chemical potential of each species.

From here:

$$Q = \frac{(C)^{\gamma}(D)^{\delta}}{(A)^{\alpha}(B)^{\beta}}$$

where, Q is the equilibrium constant of a general reaction.

The Gibbs energy change for the overall reaction is the sum of the chemical potentials of products minus those of reactants, weighted by their stoichiometric coefficients:

$$\Delta G = \sum G_{products} - \sum G_{reactants}$$

From thermodynamics, the free-energy change at arbitrary concentrations is

$$\Delta G = \Delta G^{\circ} + RT \ln Q$$

ΔG° is the standard free-energy change (all species at unit activity).
R is the gas constant (8.314 J/(mol·K).
T is the absolute temperature (K).
Q is the reaction quotient (activity of products over reactants, each raised to their stoichiometric power).

Under standard-state conditions ($Q = 1$), the electrical work relation becomes

$$\Delta G^{\circ} = -nF\Delta E^{\circ}$$

where ΔE° is the standard cell potential.

Set the two expressions for ΔG equal:

$$-nF\,\Delta E = -nf\,\Delta E^{\circ} + RT\ln Q$$

Rearrange to isolate ΔE:

$$nF\,\Delta E = nF\,\Delta E^{\circ} - RT\ln Q$$

Divide above equation by $-nF$:

$$\Delta E = \Delta E^{\circ} - RT\frac{\ln Q}{nF}$$

The above equation is known as Nernst equation (1887) that works for dilute solutions.

Biological reactions take place at pH around 7. At this pH 7, the standard half-cell potential is denoted by $E^{\circ\prime}$. To correct the E° values, the Nernst equation is employed:

$$E^{\circ\prime} = E^{\circ} - RT\frac{\ln Q}{nF}$$

The ° sign is not omitted, because the conditions are still standard (T, P) except concentrations.

Practice: Calculate the standard half-cell potential of the half-reaction for the reduction of gaseous oxygen (O_2) to liquid water (H_2O) for which E° is $+\ 1.229$ V.

Solution:

Half-reaction for the reduction of O_2:

$$O_2(g) + 4H + (aq) + 4e - 2H_2O(l)$$

$$\begin{aligned} E^{0\prime} &= +1.229V - \frac{8.314 \times 298}{4 \times 96845}\ln\frac{H_2O}{(O_2)(H^{4+})} \\ &= 1.229V - 0.006\ln\frac{1^2}{1 \times (10^{-7})^4} \\ &= 1.229 - 0.4123 = 0.816V \end{aligned}$$

Practice: If ΔG of glucose oxidation by oxygen under standard conditions is – 2878 kJ/mol, what would be ΔG under standard conditions and pH 7?

Solution: The value of $E^{\circ'}$ is taken from Standard Table (not provided in this book).

$$6CO_2(g) + 24H + (aq) + 24e^- \rightarrow C_6H_{12}O_6 + 6H_2O \quad [\times - 1] \quad E0' = -0.43V$$

$$O_2(g) + 4H + (aq) + 4e^- \rightarrow 2H_2O(l) \quad [\times 6] \quad E0' = +0.816V$$

$$C_6H_{12}O_6 + 6O_2(g) \rightarrow 6H_2O(l) + 6CO_2(g) \quad \Delta E0' = 0.816V - (-0.43V) = 1.246V$$

$$\Delta G^{0'} = -nF\Delta E^{0'} = -24\frac{\text{mol e}^-}{\text{mol Glucose}} \times 96485\frac{\text{C}}{\text{mole}^-} \times 1 \cdot 246\frac{\text{J}}{\text{C}}$$

$$= -2885\frac{\text{kJ}}{\text{mol Glucose}}$$

Rittmann and McCarty (2001) outlined the procedures for formulating half-reactions and determining their standard reduction free energies ($\Delta G^{\circ'}$, in kJ per electron equivalent). Half-reactions for substrate oxidation and electron acceptance are first paired to define an energy-yielding reaction, whose free-energy change (ΔG_r) reflects the Gibbs energy released during catabolism. In parallel, a half-reaction linking the same electron donor to cell-constituent precursors yields the biosynthesis reaction, with its own free-energy requirement (ΔG_s). To derive the overall stoichiometry for microbial growth, these two reactions are then blended in a ratio that accounts for the cell's energy-conservation efficiency, ε. This mixing ratio is expressed by the coefficient which directly indicates how much of the donor is diverted to energy production versus biomass formation. So, A basically gives the ratio of the fraction of a donor associated with energy production to that associated with cell synthesis. Finally, ΔG_r itself is calculated from the standard reduction potentials of the donor and acceptor half-reactions—corrected for the actual activities and stoichiometric coefficients of all species involved—ensuring that the overall growth equation is both mass- and energy-balanced.

The true or maximum cell yield (Y) is determined from A that can be expressed in electron equivalent (e^-eq) units as the fraction of donor electron equivalents converted for synthesis f_s^0, while the remaining fraction is used for energy (f_e^0) (McCarty 2007):

$$A = \frac{f_e^0}{f_s^0}$$

$$f_s^0 = \frac{1}{1+A}$$

$$f_e^0 = \frac{A}{1+A}$$

$$f_e^0 + f_s^0 = 1$$

2.9.1 *Thermodynamic Electron Equivalents Model (TEEM) to Obtain Microbial-Metabolic Reactions*

Formulating fully balanced metabolic reactions by hand is a tedious and error-prone process for biologists, ecologists, modelers, and engineers alike. To address this, a new open-source application—MbT-Tool (Metabolism based on Thermodynamics)—has been developed in NetLogo to automate the generation of Microbial Metabolic Reactions (MMRs). By leveraging the Thermodynamic Electron Equivalents Model (TEEM), MbT-Tool can rapidly assemble stoichiometrically balanced reaction equations for a variety of microbial guilds—including aerobic heterotrophs, nitrifiers, denitrifiers, methanogens, sulfate reducers, sulfide oxidizers, and fermenters (Granda et al. 2016). Figure 2.4 presents a schematic of the MbT-Tool workflow, illustrating how TEEM principles—particularly the integration of catabolic electron flows with anabolic synthesis demands—are used to derive each MMR.

This developed computational tool can be downloaded from: https://ccl.northwestern.edu/netlogo/download.shtml (to install NetLogo), and http://mosimbio.upc.edu/en/publications/publications-by-year/years-2010-now (to download the MbT-Tool).

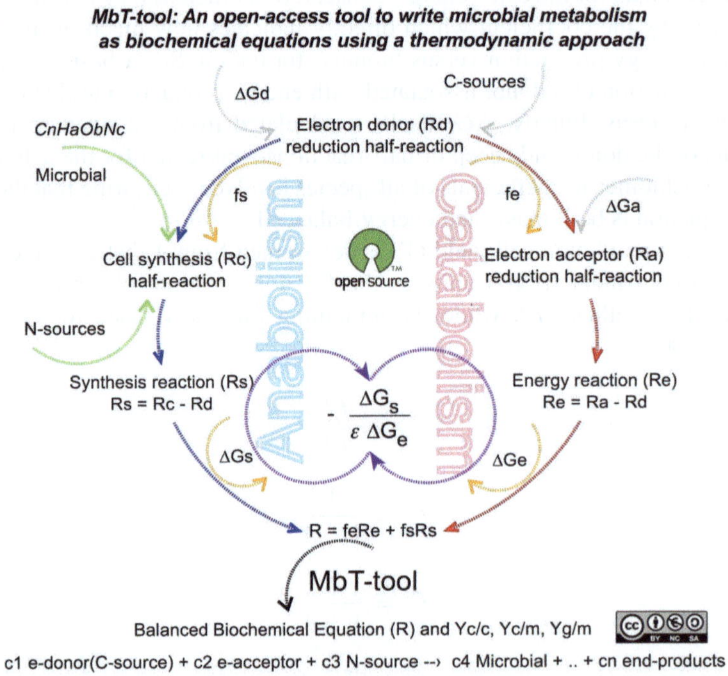

Fig. 2.4 MbT-Tool is a computational tool to represent the microbial metabolism through a microbial metabolic reaction (R) using TEEM as thermodynamic approach which is based on the energy-transfer-efficiency between catabolic and anabolic processes (Granda et al. 2016)

Further Readings

Granda PA, Gras A, Ginovart M (2016) MbT-tool: an open-access tool based on thermodynamic electron equivalents model to obtain microbial-metabolic reactions to be used in biotechnological process. Comput Struct Biotechnol J 14:325–332

Madigan MT, Martinko JM, Parker J (1997) Brock biology of microorganisms, 8th edn. Prentice Hall International Inc., New York

McCarty PL (1975) Stoichiometry of biological reactions. Progress Water Technol 7:157–172

McCarty PL (2007) Thermodynamic electron equivalents model for bacterial yield prediction: modifications and comparative evaluations. Biotechnol Bioeng 97:377–388

Metcalf, Eddy (2003) Wastewater engineering: treatment and reuse. 4th edn. McGraw-Hill, New York

Porges N, Jasewicz L, Hoover SR (1956) Principles of biological oxidation. In: McCabe J, Eckenfelder WW (eds) Biological treatment of sewage and industrial wastes. Reinhold Publishing, New York

Rittmann BE, McCarty PL (2001) Environmental biotechnology: principles and applications. McGraw-Hill Education, New York, Chicago, San Francisco, Athens, London, Madrid, Mexico City, Milan, New Delhi, Singapore, Sydney, Toronto

Chapter 3
Oxygen Demand Metrics in Water Quality

Learning Objectives:

1. *Define and calculate BOD, COD, and ThOD for assessing organic matter.*
2. *Distinguish how each oxygen demand metric reflects water quality.*

Oxygen demand, specifically in the context of biochemical oxygen demand (BOD) and chemical oxygen demand (COD), serves as a critical measure of organic matter concentration in water and wastewater. This metric is pivotal for assessing the environmental health of aquatic ecosystems and the efficiency of wastewater treatment processes. Understanding these parameters is essential for environmental engineers, biologists, and water quality professionals.

3.1 Biochemical Oxygen Demand (BOD)

BOD quantifies the amount of dissolved oxygen needed by aerobic microorganisms to break down organic matter (basically, pollutant) present in water. It is an indirect measure of the biodegradable organic matter in water, reflecting the potential impact of wastewater discharges on the oxygen levels of receiving water bodies.

BOD test, the oldest and most common organic matter concentration parameter, is used to measure BOD. This test reflects only the biodegradable organic matter and takes 5 days; thus, value is also represented by BOD_5. After 5 days, bacterial cells begin to die due to starvation as most of the biodegradable organic matter serving as an e^- donor gets oxidized. Thereafter, oxygen consumption (BOD decrease) continues but at a much lower rate. BOD is used, often in wastewater treatment plants, as an index of the degree of organic pollution in water. In practice, it is usually expressed in milligrams O_2 per litre (Nagel et al. 1992).

Originally, the five-day timeframe used in the BOD_5 test originates from historical hydrological observations in the United Kingdom, where this duration was deemed

V. Achal, *Notes on Biological Processes in Environmental Engineering*,
https://doi.org/10.1007/978-981-95-1610-0_3

the maximum time for river water to flow from its origin to the estuary, as documented by the Great Britain Royal Commission on Sewage Disposal (1908). This standardized testing protocol later gained recognition in 1936 when the American Public Health Association's Standard Methods Committee formally adopted it as a benchmark for assessing the biodegradability of organic pollutants, chemicals, and hazardous substances in environmental monitoring practices (Jouanneau et al. 2014).

BOD can be shown in equation form as below:

$$\text{Organic matter} + O_2 \xrightarrow{\text{Microorganisms}} CO_2 + H_2O + \text{New cells} + \text{Stable products}$$

Significance of BOD in Water Quality

- **Indicator of Organic Pollution**: A high BOD value indicates a high level of organic pollution, which can lead to oxygen depletion in aquatic environments. This depletion can harm aquatic life, leading to fish kills and the disruption of aquatic ecosystems.
- **Wastewater Treatment Monitoring**: BOD is a standard regulatory and operational tool used to gauge the effectiveness of wastewater treatment processes. It helps in determining the organic load removed by the treatment and the potential impact of the effluent on the receiving environment.
- **Environmental Compliance**: Many regulatory frameworks set limits on the BOD levels of wastewater discharges to protect water quality. Monitoring BOD is thus critical for compliance with environmental regulations.

3.2 Chemical Oxygen Demand (COD)

While BOD focuses on biodegradable organic matter, COD measures the total quantity of oxidizable substances (including both biodegradable and non-biodegradable organic matter) in the water, providing a more comprehensive picture of the water's organic content. COD is defined as the amount of oxygen required to chemically oxidize all the organic matter in water/wastewater by strong oxidant. This test is performed under acidic and high temperature conditions.

The most commonly used oxidant is potassium dichromate ($K_2Cr_2O_7$, which is used in combination with boiling **sulphuric acid**, where Cr^{6+} is reduced to Cr^{3+} and oxygen demand is translated according to the number of electrons transferred. Compared to BOD, COD is fast test and takes 2 h. COD is always bigger than BOD as it takes all organic matter in account.

Equation-wise, it can be presented as below:

$$\text{Organic matter} + Cr^{6+} \rightarrow Cr^{3+} + CO_2 + H_2O$$

Potassium dichromate ($K_2Cr_2O_7$) oxidation has been the internationally standardized method for assessing chemical oxygen demand (COD) since the late 1970s

(Kolb et al. 2017). However, many water samples—including municipal and industrial wastewater, as well as seawater—contain high chloride concentrations (often in grams per liter). Since chloride is also oxidized by hexavalent chromium [Cr(*VI*)], this interference produces artificially high and non-reproducible COD results. To mitigate this issue, mercury sulfate ($HgSO_4$) is added to mask chloride ions through complexation. Additionally, silver sulfate (Ag_2SO_4) in sulfuric acid serves as a catalyst to enhance the oxidation of persistent organic compounds (Moore et al. 1951).

Advantages of COD

- **Broader Scope**: COD encompasses all organic compounds, offering a fuller assessment of water pollution, including substances that are not readily biodegradable.
- **Faster Results**: Unlike the BOD test, which takes five days, COD can be measured within a few hours, making it a quicker indicator of water quality.

3.3 Comparative Insights

- **BOD versus COD**: BOD is more specific to biologically available pollution, whereas COD provides a broader measure of total organic pollution. The ratio of BOD to COD can indicate the biodegradability of organic matter in the water. The comparison between BOD and COD tests is summarized in Table 3.1.
- **Application Considerations**: The choice between BOD and COD testing depends on the specific requirements of water quality assessment or treatment process control. BOD is preferred for assessing the biological impact on natural waters and for biological treatment processes. In contrast, COD is useful for industrial applications where chemical pollutants are of concern or when rapid results are needed.

3.4 ThOD (Theoretical Oxygen Demand)

Theoretical Oxygen Demand (ThOD) is an important concept in the field of chemical and environmental engineering, especially in the context of water and wastewater treatment. ThOD is calculated based on the stoichiometric oxidation of organic compounds found in water or wastewater. This calculation involves determining the amount of oxygen required to completely oxidize the organic substances to end products like carbon dioxide (CO_2), ammonia (NH_3), phosphate ($H_2PO_4^-$), sulfate (SO_4^{2-}), and water (H_2O) (Baker et al. 1999). The calculation of ThOD is crucial for several reasons:

Table 3.1 Comparison of BOD and COD tests

Feature	BOD test	COD test
Definition	Measures the amount of oxygen required by aerobic microorganisms to decompose organic matter in water	Measures the amount of oxygen equivalent of the organic and inorganic matter in water that can be oxidized by a chemical oxidant
Purpose	Used to estimate the biodegradable organic matter content of water	Used to estimate the total quantity of organic and inorganic pollutants in water
Time required	Typically, 5 days (BOD_5) at 20 °C	Usually a few hours
Oxidant used	Oxidation by microorganisms	$K_2Cr_2O_7$ $Mn_2(SO_4)_3$
Sensitivity	Sensitive to biodegradable organic matter	Sensitive to both biodegradable and non-biodegradable organic matter
Typical use	Assessing the effectiveness of wastewater treatment plants in removing biodegradable organic pollution	Quick evaluation of water quality and determination of pollution load
Indicator of	Biological impact on aquatic life due to organic pollution	Overall oxidation capacity of pollutants, including chemical substances that may not be biologically degradable
Interference	Can be affected by the presence of toxic substances that inhibit microbial activity	Less susceptible to interference from toxic substances due to the chemical nature of the oxidation
Results interpretation	Lower BOD indicates less biodegradable organic matter and potentially less impact on aquatic life	Higher COD values indicate a higher concentration of oxidizable substances, suggesting higher levels of pollution
Regulatory relevance	Commonly used in environmental regulations to set discharge limits for sewage and wastewater	Often used for industrial wastewater monitoring and compliance with environmental regulations

- **Predictive Analysis**: It allows for a predictive analysis of the oxygen demand of various organic compounds before they are introduced into a water body or treatment system. This can help in the design and optimization of treatment processes to ensure adequate removal of organic pollutants.
- **Environmental Impact Assessment**: ThOD can be used to estimate the potential impact of organic pollutants on the oxygen levels of aquatic environments. By understanding the theoretical oxygen demand, environmental engineers can predict the possible effects of wastewater discharges on the dissolved oxygen balance of receiving waters.

- **Comparison with Empirical Measures**: ThOD provides a theoretical benchmark against which empirical measurements like COD and BOD can be compared. This comparison can help in assessing the efficiency of treatment processes and the biodegradability of organic matter in water.

3.4.1 COD Estimation from ThOD

The relationship between COD and ThOD is defined by the equation:

$$\text{COD} = \alpha \times \text{ThOD}$$

where α is an empirical constant that typically ranges between 0.95 and 1.0. This constant accounts for the efficiency of the chemical oxidation process in converting organic matter to its fully oxidized products under the conditions of the COD test.

Empirical Constant (α): The value of α reflects the fraction of the theoretical oxygen demand that is actually measured by the COD test. A value of 1.0 would indicate that the COD test captures the complete theoretical oxygen demand, while values less than 1.0 reflect the practical limitations of the COD test in fully oxidizing all organic compounds present in the sample.

3.4.2 Applications and Implications

- **Treatment Process Design and Optimization**: Understanding ThOD and its relation to COD can assist engineers in designing and optimizing wastewater treatment processes to effectively remove organic pollutants and reduce oxygen demand.
- **Regulatory Compliance**: ThOD calculations can support compliance with environmental regulations by providing theoretical estimates of oxygen demand for various pollutants, aiding in the development of discharge permits and treatment standards.
- **Research and Development**: In the research context, ThOD is valuable for studying the oxidation kinetics of organic pollutants and for developing more efficient treatment technologies that can better target and remove these pollutants from water.

In conclusion, ThOD serves as a theoretical foundation for estimating the oxygen demand of organic matter in water and wastewater treatment contexts. By providing a stoichiometric basis for understanding oxygen demand, ThOD enhances the ability of engineers and environmental professionals to predict, manage, and treat organic pollution in aquatic environments.

Practice: For a wastewater containing 100 g/L $C_2H_{12}O_6$, calculate the concentration (g/L) of oxygen required to completely oxidize the organic matter.

Solution: This question basically demands calculating ThOD, where organic matter will react with oxygen as below.

$$C_2H_{12}O_6 + 2O_2 \rightarrow 2CO_2 + 6H_2O$$

To calculate the concentration of oxygen required to completely oxidize the organic matter ($C_2H_{12}O_6$) in wastewater, we need to write the balanced chemical equation for the oxidation reaction. The general formula for the complete oxidation of an organic compound ($C_2H_{12}O_6$) to carbon dioxide (CO_2) and water (H_2O) is:

$$C_2H_{12}O_6 + O_2 \rightarrow CO_2 + H_2O$$

First, we balance this chemical equation:

$$C_2H_{12}O_6 + O_2 \rightarrow 2CO_2 + 6H_2O$$

To balance the oxygen, we have:

$$C_2H_{12}O_6 + 2O_2 \rightarrow 2CO_2 + 6H_2O$$

So, one mole of $C_2H_{12}O_6$ (which has a molar mass of approximately 2(12) + 12(1) + 6(16) = 180 g/mol) requires 6 mol of O_2 to be completely oxidized. The molar mass of O_2 is approximately 32 g/mol.

Given that the wastewater contains 100 g/L of $C_2H_{12}O_6$, we calculate the concentration of oxygen required (in g/L) using the stoichiometry of the reaction:

$$\text{Oxygen required} = \frac{100\frac{g}{L}C_2H_{12}O_6}{180\frac{g}{mol}C_2H_{12}O_6} \times 6\,\text{moles}\,O_2 \times 32\frac{g}{mol}O_2$$

Let's calculate this:

The concentration of oxygen required to completely oxidize the organic matter ($C_2H_{12}O_6$) in the wastewater is approximately 106.67 g/L.

Practice: During a COD test of wastewater sample, the concentration of $K_2Cr_2O_7$ decreased from 500 to 50 mg/L. Calculate the COD of the sample?

Solution:

Step 1: Convert the change in concentration of $K_2Cr_2O_7$:

Initial concentration of $K_2Cr_2O_7$: 500 mg/L.

Final concentration of $K_2Cr_2O_7$: 50 mg/L.

Change in concentration ($\Delta concentration$):$500\,\text{mg/L} - 50\,\text{mg/L} = 450\,\text{mg/L}$.

Step 2: Convert the change in concentration to moles of $K_2Cr_2O_7$ reacted:

The molar mass of $K_2Cr_2O_7$ is 294.185 g/mol.

Moles of $K_2Cr_2O_7$ reacted = (450 mg/L/294.185 g/mol) × (1 g/1000 mg) = 0.0015 mol/L

$$\text{Moles of } K_2Cr_2O_7 \text{ reacted} = \frac{450\frac{\text{mg}}{\text{L}}}{294.185\frac{\text{g}}{\text{mol}}} \times \frac{1\,\text{g}}{1000\,\text{mg}} = 0.0015\frac{\text{mol}}{\text{L}}$$

Step 3: Now calculate COD:

$\text{COD}(\text{mgO}_2/\text{L}) = \text{Moles of } O_2 \text{ equivalent} \times 32\,\text{g/mol} \times 1000\,\text{mg/g} = 48\,\text{mgO}_2/\text{L}$

Further Readings

Baker JR, Milke MW, Mihelcic JR (1999) Relationship between chemical and theoretical oxygen demand for specific classes of organic chemicals. Water Res 33:327–334

Great Britain, Royal commission on sewage disposal, 1908. Final report of the commissioners appointed to inquire and report what methods of treating and disposing of sewage (including any liquid from any factory or manufacturing process) may properly be adopted. General summary of conclusions and recommendations, Cornell University Library ed

Jouanneau S, Recoules L, Durand MJ, Boukabache A, Picot V, Primault Y, Lakel A, Sengelin M, Barillon B, Thouand G (2014) Methods for assessing biochemical oxygen demand (BOD): a review. Water Res 49:62–82

Kolb M, Bahadir M, Teichgräber B (2017) Determination of chemical oxygen demand (COD) using an alternative wet chemical method free of mercury and dichromate. Water Res 122:645–654

Moore WA, Ludzack FJ, Ruchhoft CC (1951) Determination of oxygen-consumed values of organic wastes. Anal Chem 23(9):1297–1300

Nagel B, Dellweg H, Gierasch LM (1992) Glossary for chemists of terms used in biotechnology (IUPAC Recommendations 1992). Pure Appl Chem 64:143–168

Chapter 4
Catabolism

Learning Objectives:

1. *Explain the ATP-producing catabolic pathways (glycolysis, Krebs cycle, oxidative phosphorylation) and calculate ATP yield from NADH and $FADH_2$.*
2. *Describe how bacterial catabolism sustains cellular growth and maintenance by converting substrates into usable energy.*

Understanding metabolism is essential to grasp the intricacies of how cells, including those of bacteria, function at a biochemical level. Metabolism encompasses the entirety of biochemical reactions within a cell or organism, forming the foundation for life's processes. Bacterial metabolism, in particular, involves the study of how these microorganisms take in and utilize various inorganic or organic compounds to sustain growth and maintain a balanced cellular environment.

4.1 Key Functions of Metabolic Reactions

Metabolic reactions within bacteria serve two primary purposes:

- **Energy Production**: These reactions are crucial for generating the energy required by the cell to carry out various functions such as growth, reproduction, movement, and the maintenance of cellular structures. This energy production is primarily achieved through the breakdown of compounds, a process known as catabolism.
- **Biosynthesis and Assimilation**: Metabolic reactions are also responsible for the synthesis of new cellular materials. This includes the creation of nucleic acids, proteins, lipids, and carbohydrates that are essential for cell structure, function, and regulation. The assimilation of these materials enables the cell to repair, grow, and reproduce.

V. Achal, *Notes on Biological Processes in Environmental Engineering*,
https://doi.org/10.1007/978-981-95-1610-0_4

4.2 Catabolism and Anabolism: The Two Facets of Metabolism

Metabolism is divided into two interconnected processes: catabolism and anabolism.

Catabolism involves the breakdown of complex organic molecules into simpler ones, releasing energy in the process. This energy is then captured in the form of ATP (adenosine triphosphate), which can be used by the cell for various activities. Catabolic reactions are essential for removing old cellular components and providing the raw materials and energy for anabolism.

Anabolism is the process of synthesizing all complex molecules needed by the cells from simpler ones. Anabolic reactions require energy, usually derived from ATP generated during catabolic reactions, to build new cellular components. These reactions are crucial for cell growth, reproduction, and repair.

The Interconnected Nature of Metabolism

While it is convenient to study catabolism and anabolism as separate processes, they are inherently interconnected within the cell. The energy produced by catabolic reactions fuels anabolic processes, and the breakdown of compounds provides the necessary substrates for biosynthesis (Fig. 4.1). This dynamic interplay ensures that cells can efficiently respond to environmental changes, maintain homeostasis, and support life's complexity.

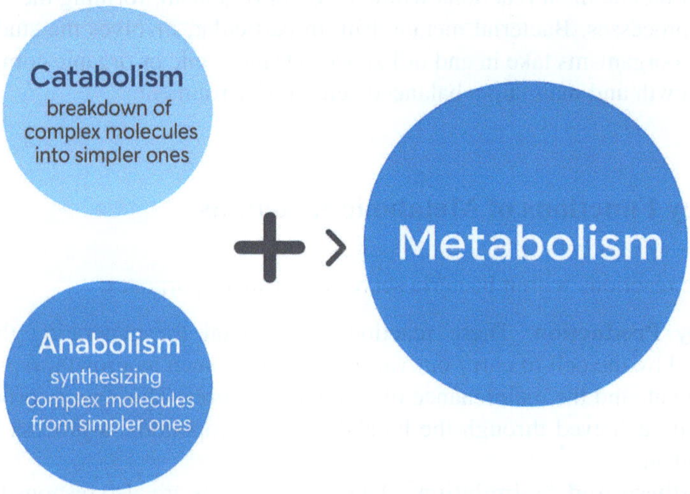

Fig. 4.1 Metabolism—simplified

4.3 Focusing on Catabolism

Understanding catabolism is fundamental to exploring bacterial metabolism, given its central role in energy production and provision of substrates necessary for biosynthesis. Catabolic processes involve the systematic breakdown of complex organic molecules into simpler, energy-rich intermediates, highlighting the adaptability and metabolic versatility inherent in microbial life. By examining catabolism, we gain deeper insights into the intricate biochemical networks that enable bacterial growth, maintenance, and survival. Furthermore, these insights have direct practical implications in environmental biotechnology applications such as wastewater treatment and bioremediation.

Although catabolism and anabolism are intrinsically intertwined and collectively constitute the broader metabolic network of the cell (Fig. 4.2), it is practical for clarity to address these processes separately.

The primary objective of wastewater treatment is the reduction of organic and energy-rich pollutants present in water. This reduction is principally achieved through microbial catabolism. During wastewater treatment, bacterial populations utilize diverse substrates—carbohydrates, lipids, proteins, and other macromolecules—as sources of electrons, carbon, and energy. As these macromolecules undergo microbial degradation, they are transformed into simpler molecules, releasing energy in a controlled, stepwise oxidation process. This energy is captured in intermediate metabolites and conserved through electron carriers, primarily nicotinamide adenine dinucleotide (NADH), and stored in the universal energy currency, adenosine triphosphate (ATP). The ATP and NADH generated during catabolic pathways subsequently supply energy to various cellular functions and biosynthetic processes.

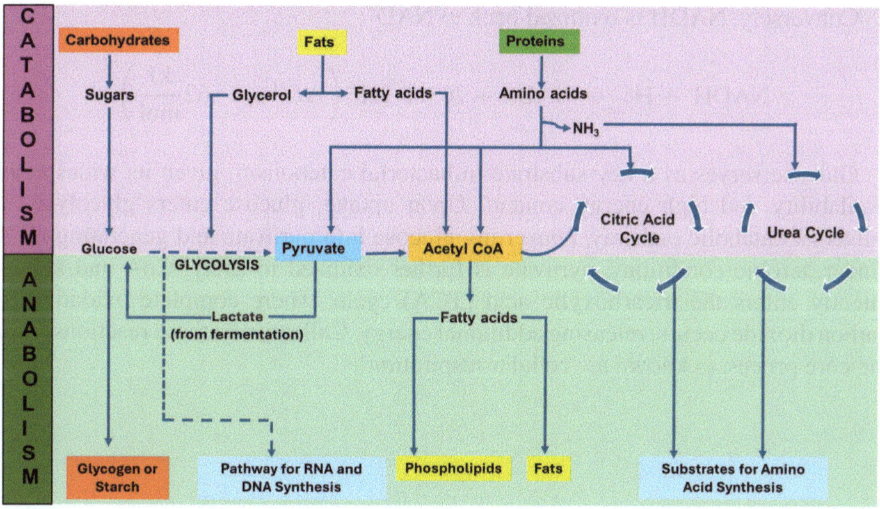

Fig. 4.2 The interconnected processes of catabolism and anabolism

Adenosine 5′-triphosphate (ATP)

ATP is the primary energy carrier molecule across all living organisms. Structurally, ATP functions as a coenzyme, transferring energy within cells through the hydrolysis of its phosphate groups. The outermost phosphate group (Pi, inorganic phosphate) is cleaved, converting ATP into adenosine diphosphate (ADP), thus releasing energy essential for cellular work:

$$\mathrm{ATP} + \mathrm{H_2O} \rightarrow \mathrm{ADP} + \mathrm{Pi}\left(\Delta G^{0'} = -36\frac{\mathrm{kJ}}{\mathrm{mol}}\right)$$

Conversely, energy released from catabolic reactions can regenerate ATP from ADP:

$$\mathrm{ADP} + \mathrm{Pi} \rightarrow \mathrm{ATP} + \mathrm{H_2O}\left(\Delta G^{0'} = +36\frac{\mathrm{kJ}}{\mathrm{mol}}\right)$$

This cycle underscores ATP's critical role as the cellular "energy currency."

Nicotinamide adenine dinucleotide (NAD^+/NADH)

NAD^+ is another vital coenzyme involved in cellular energy metabolism. It acts primarily as an electron carrier, accepting electrons from catabolic reactions to form NADH, which subsequently contributes electrons to the electron transport chain, driving ATP synthesis:

$$\mathrm{NAD^+} + 2\mathrm{e^-} + 2\mathrm{H^+} \rightarrow \mathrm{NADH} + \mathrm{H^+}\left(\Delta G^{0'} = 62\frac{\mathrm{kJ}}{\mathrm{mol}}\right)$$

Conversely, NADH is oxidized back to NAD^+:

$$\mathrm{NADH} + \mathrm{H^+} \rightarrow \mathrm{NAD^+} + 2\mathrm{e^-} + 2\mathrm{H^+}\left(\Delta G^{0'} = -62\frac{\mathrm{kJ}}{\mathrm{mol}}\right)$$

Glucose serves as a key substrate in bacterial catabolism, given its widespread availability and high energy content. Upon uptake, glucose enters glycolysis, a universal catabolic pathway, converting glucose into pyruvate and generating ATP. Under aerobic conditions, pyruvate is further oxidized to acetyl-CoA and subsequently enters the tricarboxylic acid (TCA) cycle, where complete oxidation to carbon dioxide occurs, releasing additional energy. Collectively, these reactions form the core processes known as 'cellular respiration'.

4.4 Cellular Respiration

Cellular respiration is the primary mechanism by which cells extract energy from nutrient substrates, primarily glucose, converting it efficiently into ATP. It encompasses four distinct yet interconnected stages, in sequence as follow:

- Glycolysis
- Pyruvate oxidation
- Citric acid cycle
- Oxidative phosphorylation

Glycolysis occurs in the cytoplasm, where glucose (a six-carbon sugar) is enzymatically broken down into two molecules of pyruvate (a three-carbon compound). This step yields a small amount of ATP and NADH.

Pyruvate oxidation bridges glycolysis and the citric acid cycle, converting pyruvate into acetyl-CoA, generating NADH, and releasing carbon dioxide as a by-product.

In the citric acid cycle, acetyl-CoA is fully oxidized into carbon dioxide. This stage captures high-energy electrons in NADH and flavin adenine dinucleotide ($FADH_2$), both essential for the subsequent step.

Oxidative phosphorylation, the final stage occurring in the cell membrane in bacteria, utilizes electrons from NADH and $FADH_2$, transferring them through a chain of membrane-bound electron carriers. This electron transport generates a proton gradient across the membrane, driving ATP synthesis via ATP synthase. This step accounts for the majority of ATP produced in cellular respiration.

By elucidating these pathways, we appreciate the efficiency of bacterial energy metabolism and their capacity to thrive in diverse environmental niches, including engineered systems for wastewater treatment and natural ecosystems for bioremediation.

4.5 Glycolysis (G. glyco = sugar, lysis = breakdown)

Glycolysis takes place in cytoplasm. It exists in all organisms. In this process, glucose—a six-carbon sugar—undergoes a series of chemical transformations mediated by bacterial enzymes gets converted into two molecules of pyruvate, a three-carbon organic molecule (Fig. 4.3). It is composed of ten enzyme-catalyzed steps.

The first five steps of glycolysis are regarded as the preparatory (energy requiring) phase, since two molecules of ATP are invested to convert the glucose into two three-carbon sugar phosphates. The second half (last five steps) of glycolysis is known as the payoff (energy releasing) phase, characterized by a generation of four molecules of ATP and two molecules of NADH.

- **Preparatory phase**: Glucose $\rightarrow$ fructose-1,6-bisphosphate, consuming 2 ATP.

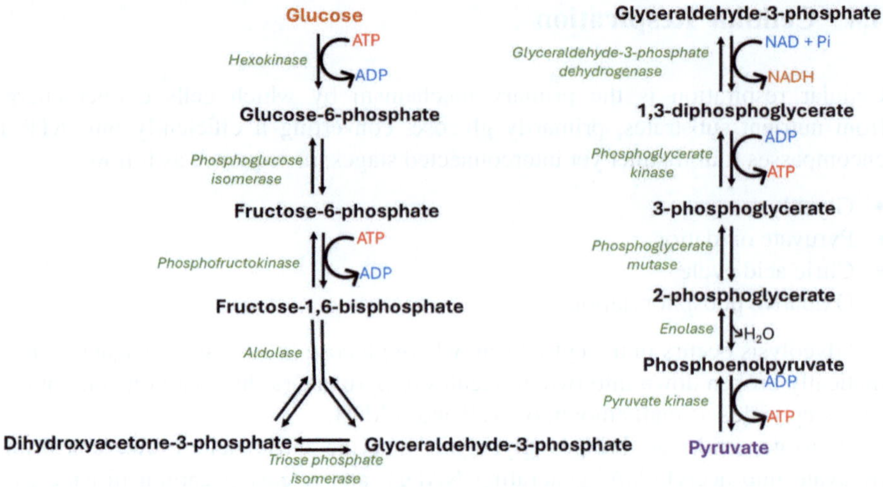

Fig. 4.3 Overview of glycolysis: preparatory phase (left) and payoff phase (right). The payoff phase reactions occur twice since one molecule of six-carbon glucose splits into two molecules of three-carbon glyceraldehyde-3-phosphate

- **Payoff phase**: Glyceraldehyde-3-phosphate → pyruvate, producing 4 ATP and 2 NADH.

  Glycolysis reaction:

$$\text{Glucose} + 2\,\text{ATP} + 2\,\text{NAD}^{+} + 4\,\text{ADP} + 2\,\text{Pi}$$
$$\rightarrow 2\,\text{Pyruvate} + 2\,\text{ADP} + 2\,\text{NADH} + 2\text{H}^{+} + 4\,\text{ATP} + 2\text{H}_2\text{O}$$

Although 4 ATPs are produced, the net gain of ATP in anaerobic glycolysis of glucose is only two. This is because the two ATPs are used during early stages of Glycolysis. The net equation is shown below:

$$\text{Glucose} + 2\text{NAD}^{+} + 2\text{ADP} + 2\text{Pi} \rightarrow 2\text{Pyruvate} + 2\text{NADH} + 2\text{H}^{+} + 2\text{ATP} + 2\text{H}_2\text{O}$$

End product of glycolysis: 2Pyruvate + 2ATP + 2NADH.

4.6 Fermentation

The end product of glycolysis is pyruvate, which in the absence of oxygen enters into fermentation. In other words, glycolysis can take place without oxygen in a process called fermentation. Ethanol fermentation (Fig. 4.4) and lactic acid fermentation (Fig. 4.5) are common examples, carried out by bacteria including *Zymomonas mobilis* and *Lactobacillus*, respectively.

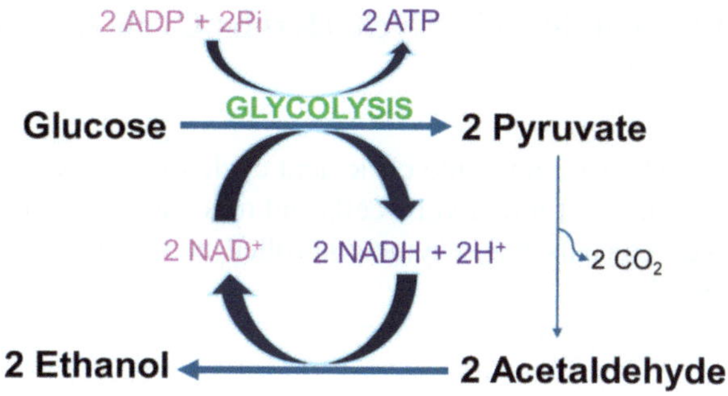

Fig. 4.4 Ethanol fermentation

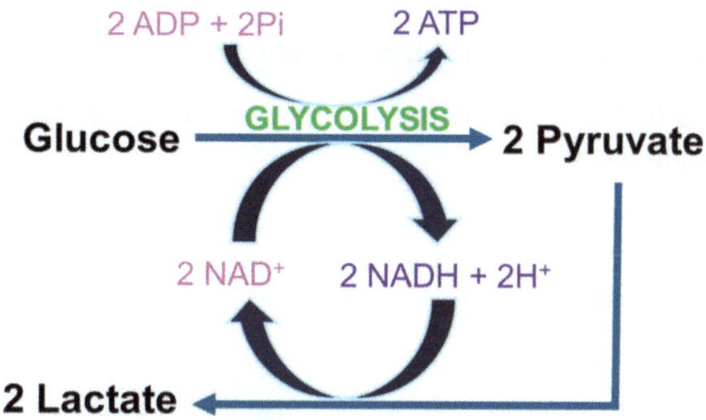

Fig. 4.5 Lactic acid fermentation

Ethanol fermentation doesn't occur in human body because humans don't produce pyruvate decarboxylase.

Fermentation yields no additional ATP but sustains glycolysis by recycling NADH to NAD^+.

4.7 Pyruvate Oxidation (Oxidative Decarboxylation of Pyruvate)

In the presence of oxygen, pyruvate enters into mitochondria where it is converted to acetyl coA, catalyzed by pyruvate dehydrogenase complex.

$$\text{Pyruvate } + \text{CoA} - \text{SH} + \text{NAD}^+ \rightarrow \text{Acetyl} - \text{CoA} + \text{CO}_2 + \text{NADH} + \text{H}^+$$

Final product from one molecule of glucose: 2 acetyl CoA.

Products of pyruvate oxidation: $(1CO_2 + 1NADH) \times 2$.

4.8 Citric Acid Cycle (Krebs Cycle or Tricarboxylic Acid, TCA, Cycle)

Acetyl coA enters into citric acid cycle that is a series of reaction take place in the cytoplasm of prokaryotic cells and mitochondria of eukaryotic cells. The citric acid cycle consists of **8** enzyme-controlled steps (Fig. 4.6). It is a central driver of cellular respiration.

Products of Citric acid cycle:

$$(1\text{ATP} + 2\text{CO}_2 + 3\text{NADH} + 1\text{FADH}_2) \times 2$$

Key players in the citric acid cycle:

ATP: The basic energy currency of the cell. It's a form of energy that cells can use right away.

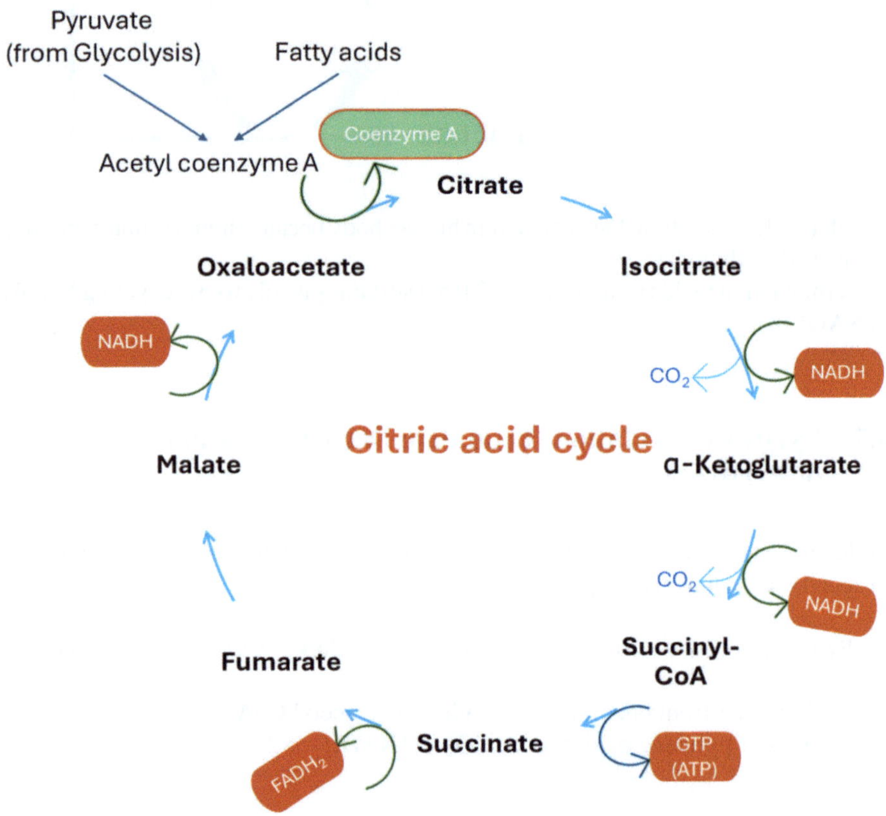

Fig. 4.6 The citric acid cycle

GTP: Similar to ATP, GTP can be easily converted to ATP in the cell.

NADH:: An energy shuttle which delivers high energy electrons to the electron transport chain (discussed later) where they will eventually power the production of 2–3 ATP molecules.

FADH $_2$: Another energy shuttle that carries high energy electrons to the electron transport chain, where they will ultimately drive production of 1 to 2 ATP molecules.

4.9 Oxidative Phosphorylation

The landmark discovery in 1948 by Eugene Kennedy and Albert Lehninger that mitochondria serve as the site of oxidative phosphorylation in eukaryotic cells initiated the modern era of research in biological energy transduction (Kennedy and Lehninger 1949). Oxidative phosphorylation produces most of the ATP made in cellular respiration and acts as ATP factory. ATPs produced in glycolysis and citric acid cycle give energy directly but what about NADH and $FADH_2$?

Oxidative phosphorylation is a redox reaction, in which NADH and/$FADH_2$ generated during glycolysis, pyruvate oxidation and citric acid cycle gives electrons to the terminal e- acceptor. The products are NAD^+ and/FAD and reduced e-acceptor (for example, H_2O, N_2, H_2S).

There are two processes involved in oxidative phosphorylation, one to pass electrons during NADH or $FADH_2$ oxidation via electron transport chain (ETC) and another to make ATP (chemiosmosis).

The electron transport chain (ETC) is a series of proteins and organic molecules, organized into four large complexes labeled I to IV, found in the inner membrane of the mitochondria. Electrons are passed from one member of the transport chain to another in a series of redox reactions. Energy released in these reactions is captured as a proton gradient, which is then used to make ATP in a process, called chemiosmosis. Together the ETC and chemiosmosis make up oxidative phosphorylation (Fig. 4.7).

There are four key steps of oxidative phosphorylation:

- *Delivery of Electrons by NADH and* $FADH_2$

Reduced NADH and $FADH_2$ formed during other steps of cellular respiration transfer their electrons to molecules near the beginning of the transport chain. After transferring the electrons, they get oxidized to NAD^+ and FAD and are utilized in other steps of cellular respiration.

- *Electron Transport and Proton Pumping*

The electrons move from a higher energy level to a lower energy level, thereby releasing energy. Some of the energy is used to move the electrons from the mitochondrial matrix to the intermembrane space. Thus, an electrochemical gradient is established.

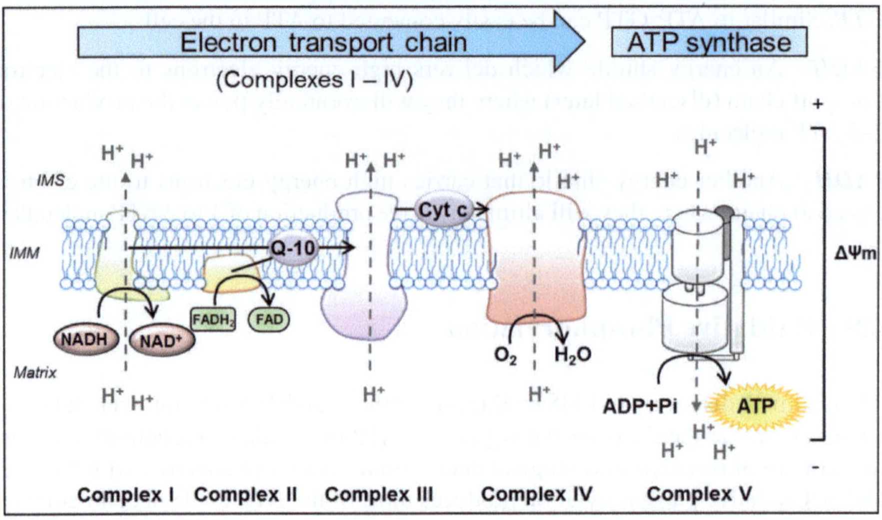

Fig. 4.7 Schematic representation of oxidative phosphorylation (OXPHOS) in mitochondria. The process of OXPHOS is the main pathway in the cell to produce adenosine triphosphate (ATP). It consists of two coupled processes embedded in the inner mitochondrial membrane (IMM), the electron transport chain (ETC) and the ATP synthesis. In the ETC, the reduced substrates NADH and $FADH_2$ are oxidized by the NADH-ubiquinone oxidoreductase (Complex I) and succinate-CoenzymeQ reductase (Complex II), respectively. These proteins transfer electrons from their substrates onto Q-10, which serves as a substrate for the CoenzymeQ-cytochrome c oxidoreductase (Complex III). Q-10 is a highly lipophilic substance that is able to diffuse within the IMM. Complex III transfers the electrons from Q-10 onto cytochrome c, which is a water-soluble electron carrier, located at the surface of the IMM in the IMS. In the final step of the ETC, cytochrome c oxidase (Complex IV) uses the electrons from reduced Cyt c to reduce molecular oxygen to water. In the process of transferring electrons, the Complexes I, III and IV actively move protons (H^+) from the mitochondrial matrix to the IMS, forming the $\Delta\Psi$m. Ultimately, this potential is used by the ATP synthase (Complex V) to catalyze the generation of ATP from ADP and Pi. ADP: adenosine diphosphate, ATP: adenosine triphosphate, Cyt c: cytochrome c, ETC: electron transport chain, $FADH_2$: flavin adenine dinucleotide, IMM: inner mitochondrial membrane, IMS: intermembrane space, NADH: nicotinamide adenine dinucleotide, Pi: inorganic phosphate, Q-10: coenzymeQ10, $\Delta\Psi$m: mitochondrial membrane potential (Szabo et al. 2020)

- *Splitting of Oxygen to form Water*

At the end of ETC, the electrons are then transferred to the molecular oxygen, which splits into half and uptakes H^+ to form water.

- *ATP Synthesis*

The transport chain builds a proton gradient across the inner mitochondrial membrane, with a higher concentration of H^+ in the intermembrane space and a lower concentration in the matrix. This creates proton motive gradient (waterfall of H^+). The H^+ ions pass through an enzyme called ATP synthase while flowing back into the matrix to synthesize ATP.

ATP yield

What is the number of ATP produced per glucose during cellular respiration? Depending on what you read in different books or ask different professors, you may get slightly different answers. However, most of answer varies between 30 and 38 ATP (Table 4.1).

How we do this calculation?

Before calculating the total ATP yield from cellular respiration, it is crucial to clearly understand how many ATP molecules are generated from the oxidation of electron carriers NADH and $FADH_2$ through the electron transport chain (ETC). The theoretical yield suggests that one mole of NADH produces approximately three ATP molecules, while one mole of $FADH_2$ yields about two ATP molecules (Nelson and Cox 2008).

However, experimental observations have provided a more precise and slightly lower yield. Studies show that ATP synthesis by ATP synthase requires approximately four protons (H^+) to flow back into the mitochondrial matrix through the ATP synthase complex (Complex V). This process is driven by the proton motive force established by the proton gradient across the inner mitochondrial membrane (Mitchell 1961; Walker 2013).

NADH transfers electrons to Complex I of the ETC, leading to the pumping of ten protons from the mitochondrial matrix to the intermembrane space. Since four protons are required to generate one ATP molecule, the actual experimental yield per NADH molecule is approximately 2.5 ATP (Brand 2005; Hinkle 2005).

Conversely, electrons from $FADH_2$ enter the ETC at Complex II, bypassing Complex I and resulting in fewer protons being pumped across the membrane. Specifically, $FADH_2$ oxidation leads to the translocation of only six protons. Consequently, each molecule of $FADH_2$ experimentally yields approximately 1.5 ATP molecules (Nelson and Cox 2021).

Understanding these precise proton requirements and ATP yields is essential for accurately assessing the energy efficiency of metabolic processes and has significant implications for biochemical energetics, metabolism, and bioengineering applications.

Before that it is important to know number of ATP produced from NADH and $FADH_2$. Theoretically, one mol of NADH produces 3 ATP, while one mol of $FADH_2$ generates 2 ATP. However, experimentally, it was found that four H^+ ions must flow back into the matrix through ATP synthase to power the synthesis of one ATP molecule. As NADH pumps 10 H^+ ions through ETC from matrix to inner membrane space, each NADH yields 2.5 ATP. On other hand, $FADH_2$ pumps six H^+ ions through ETC, thus one mole of $FADH_2$ yield 1.5 ATP.

Further, there are two types of shuttle system works to transfer NADH synthesized in the cytoplasm to the mitochondrial matrix so that they can enter in ETC to produce ATP. One shuttle leads to the production of 2 ATP (experimentally 1.5), while other produces 3 ATP (experimentally 2.5) per NADH.

Practice: What is the Gibbs free energy change of the following reaction:

$$\text{Glucose} + \text{Pi} \rightarrow \text{Glucose} - 6 - \text{P} + H_2O$$

How can this non-spontaneous reaction be carried out using ATP?

Solution:

Gibbs free energy change of the reaction can be taken from a standard table (not provided in this book).

$$\text{Glucose} + \text{Pi} \rightarrow \text{Glucose} - 6 - \text{P} + H_2O \quad \Delta G^{0'} = 11.6\ \text{kJ/mol}$$

$$\text{ATP} + H_2O \rightarrow \text{ADP} + \text{Pi} \quad \Delta G^{0'} = -36\ \text{KJ/mol}$$

$$\text{Glucose} + \text{ATP} \rightarrow \text{Glucose} - 6 - \text{P} + \text{ADP}$$

Table 4.1 Comparative ATP production in glucose respiration pathways (Experimental vs. Theoretical)

ATP yield (per mol of glucose)					
Stage	Direct products	Experimental ATP yield	Experimental ATP yield (due to different shuttle)	Theoretical ATP yield (due to different shuttle)	Theoretical ATP yield (maximum)
Glycolysis	2 ATP	2 ATP	2 ATP	2 ATP	2 ATP
	2 NADH	5 ATP	3 ATP	4 ATP	6 ATP
Pyruvate oxidation	2 NADH	5 ATP	5 ATP	6 ATP	6 ATP
Citric acid cycle	2 GTP	2 ATP	2 ATP	2 ATP	2 ATP
	6NADH	15 ATP	15 ATP	18 ATP	18 ATP
	2 $FADH_2$	3 ATP	3 ATP	4 ATP	4 ATP
Total ATP		32 ATP	30 ATP	36 ATP	38 ATP

$$\Delta G_0' = 11.6 + (-36) = -24.4\,\frac{\text{kJ}}{\text{mol}}$$

Practice:

(a) What is the general term used for the anaerobic degradation of glucose to obtain energy?

Answer: Fermentation.

(b) How many molecules of acetyl CoA, an acetyl group attached to 'coenzyme A', are produced from a single molecule of glucose for participation in the Krebs cycle?

Answer: Two molecules of acetyl CoA.

Further Readings

Brand MD (2005) The efficiency and plasticity of mitochondrial energy transduction. Biochem Soc Trans 33(5):897–904

Hinkle PC (2005) P/O ratios of mitochondrial oxidative phosphorylation. Biochem Biophys Acta 1706(1–2):1–11

Kennedy EP, Lehninger AL (1949) Oxidation of fatty acids and tricarboxylic acid cycle intermediates by isolated rat liver mitochondria. J Biol Chem 179(2):957–972

Mitchell P (1961) Coupling of phosphorylation to electron and hydrogen transfer by a chemiosmotic type of mechanism. Nature 191(4784):144–148

Nelson DL, Cox MM (2008) Lehninger principles of biochemistry, 5th edn. Freeman, W. H

Nelson DL, Cox MM (2021) Lehninger principles of biochemistry, 8th edn. Freeman, W. H

Szabo L, Eckert A, Grimm A (2020) Insights into disease-associated tau impact on mitochondria. Int J Mol Sci 21(17):6344

Walker JE (2013) The ATP synthase: the understood, the uncertain, and the unknown. Biochem Soc Trans 41(1):1–16

Chapter 5
Anabolism

Learning Objectives:

1. *Understand the role of anabolism in biomass synthesis during wastewater treatment.*
2. *Develop and balance stoichiometric equations that integrate catabolic and anabolic reactions, including electron and energy balances.*

5.1 Anabolism in the Context of Wastewater Treatment

The energy stored in adenosine triphosphate (ATP)—produced during catabolic reactions—acts as the primary fuel for anabolic processes. Anabolism, often referred to as biosynthesis, is the collection of biochemical reactions in which simple molecules are assembled to form more complex compounds. Unlike catabolism, which releases energy, anabolism is an endergonic process that requires an input of energy (typically in the form of ATP) to proceed.

In biological wastewater treatment, anabolism is crucial because it underpins microbial growth. Microorganisms consume the organic constituents in wastewater not only to derive energy (catabolism) but also to synthesize new cellular components (anabolism). These cellular components accumulate as biomass, enabling further degradation of pollutants. The formation of microbial biomass is thus both a tool for pollutant removal and a by-product of it.

V. Achal, *Notes on Biological Processes in Environmental Engineering*,
https://doi.org/10.1007/978-981-95-1610-0_5

5.1.1 Why Study Anabolism in Wastewater Treatment?

Understanding anabolism is vital from an engineering perspective because:

- **Biomass Growth**: The synthesis of new cells ensures that a sufficient and active microbial community is maintained, which is essential for the degradation of pollutants.
- **Process Optimization**: Predicting the rate of new cell formation is key to designing and optimizing treatment processes. Engineers rely on stoichiometric reactions for metabolism (encompassing both catabolism and anabolism) to estimate substrate requirements and biomass yields.
- **Balanced Mass Equations**: In wastewater treatment, balanced mass equations are fundamental. These equations account for the inputs—such as electron acceptors (for example, oxygen), essential nutrients (nitrogen and phosphorus), and pH adjusting chemicals (such as sulfuric acid or sodium hydroxide)—required to support microbial growth. The resulting biomass, which may eventually become excess sludge, is the end product of these carefully managed reactions.

By mastering the stoichiometric modeling of anabolic reactions, engineers can predict and control the growth kinetics of microbial populations and, consequently, optimize the removal of pollutants.

5.2 Writing Stoichiometric Reactions for Metabolism

Metabolism in wastewater treatment involves the interplay between catabolism and anabolism. Catabolism refers to the breakdown of organic substrates to release energy, while anabolism uses this energy to synthesize new cellular material. For the purpose of designing and optimizing treatment systems, it is essential to write stoichiometric equations that capture both aspects of metabolism.

5.2.1 Components of Stoichiometric Equations

When writing these equations, several key components must be considered:

- **Substrate Utilization**: This component represents the conversion of organic matter into energy and simpler compounds through catabolic reactions.
- **Energy Production and Consumption**: It is necessary to quantify the energy generated from catabolism (often expressed via electron donors and the formation of ATP) and how that energy is subsequently used in anabolic reactions.
- **Biomass Synthesis**: The synthesis of new cells requires the incorporation of key nutrients, such as nitrogen and phosphorus, into cellular macromolecules. In many models, the biomass is represented by a generic formula such as $C_5H_7O_2N$.

- **By-product Formation**: The stoichiometric equations must also account for the formation of by-products such as carbon dioxide (CO_2), water (H_2O), and, in certain systems, other gases like methane (CH_4).

5.2.2 Challenges and Simplification

The chemical composition of protoplasm—including proteins, lipids, and carbohydrates—is highly complex, and the anabolic pathways involved in synthesizing these compounds are numerous. To manage this complexity, engineers and researchers often employ simplified models:

- **Generic Cell Formula**: The common bacterial cell formula $C_5H_7O_2N$ is used to represent biomass composition.
- **Electron Partitioning**: Electrons derived from substrate oxidation are assumed to partition between energy generation (catabolism) and cell synthesis (anabolism). This partitioning is often quantified using fixed energy-to-synthesis ratios ($f_e : f_s$), which may vary depending on the treatment conditions (aerobic versus anoxic).

By applying these simplifications, stoichiometric equations can be developed that balance the elements, electrons, and energy, thus providing a robust framework for predicting metabolism in wastewater treatment systems.

5.3 The Overall Metabolic (Growth) Reaction

The overall metabolic or growth reaction is the net outcome of the combined catabolic and anabolic processes. This reaction is typically constructed in a stepwise fashion:

1. **Complete Oxidation of the Substrate**: The organic substrate is fully oxidized into CO_2 and ammonia (NH_3), releasing electrons in the process.
2. **Electron Partitioning for Energy Production**: A portion of the electrons released during oxidation is transferred to a terminal electron acceptor (typically oxygen under aerobic conditions) to generate the energy required for cellular processes. This fraction is represented by f_e.
3. **Electron Utilization for Biomass Synthesis**: The remaining electrons (represented by f_s) are used, in combination with NH_3 and CO_2, to form new cellular material (biomass).

The general strategy for writing such a metabolic reaction involves:

- Writing the individual half-reactions for reduction (pertaining to cell synthesis and for energy) and for the oxidation of the substrate.
- Balancing the electrons in these half-reactions.
- Multiplying each half-reaction by the appropriate factor (fs for cell synthesis, fe for energy generation).

- Combining the balanced reactions to obtain the overall stoichiometric equation.

Example Approach:

The reactions follow 3 steps:

(i) The substrate is completely oxidized to CO_2, and NH_3.
(ii) A portion of the electrons released in the 1st step is transferred to the terminal e- acceptor to supply energy (f_e).
(iii) A portion of the electrons released in the 1st step (f_s) reduce CO_2, and together with NH_3, synthesize new biomass.

Further, the following assumptions are important to know in writing metabolic reactions:

- Unless otherwise stated, the composition of the biomass produced in the process is $C_5H_7O_2N$.
- Unless otherwise stated, the composition of the carbon source in domestic wastewater is $C_{10}H_{19}O_3N$.
- The energy/synthesis ratio ($f_e : f_s$) for aerobic biological removal of the carbon source is 0.54:1.
- The energy/synthesis ratio ($f_e : f_s$) for anoxic biological removal of the carbon source is 1.86:1.
- The BOD_5/COD ratio of the carbon source is 0.45:1.

Organic half-reaction for domestic wastewater is written as:

$$\frac{9}{50}CO_2 + \frac{1}{50}NH_4^+ + \frac{1}{50}HCO_3^- + H^+ + e^- = \frac{1}{50}C_{10}H_{19}O_3N + \frac{9}{25}H_2O$$

Let's write a metabolic reaction assuming domestic wastewater treatment was conducted using aerobic bacteria. Remember unless otherwise stated, ammonia (NH_3) serves as nitrogen source in biomass synthesis.

Step 1: write three half-reactions.

(i) Reduction (cell synthesis): R_c

$$NH_3 + 5CO_2 + 20e^- + 20H^+ \rightarrow C_5H_7O_2N + 8H_2O(1) \times f_s \tag{5.1}$$

(ii) Reduction (energy): R_a

$$O_2 + 4e^- + 4H^+ 2H_2O(2) \times f_e \tag{5.2}$$

(iii) Oxidation of substrate in wastewater: $-R_d$

$$C_{10}H_{19}O_3N + 17H_2O 10CO_2 + NH_3 + 50e^- + 50H^+(3) \tag{5.3}$$

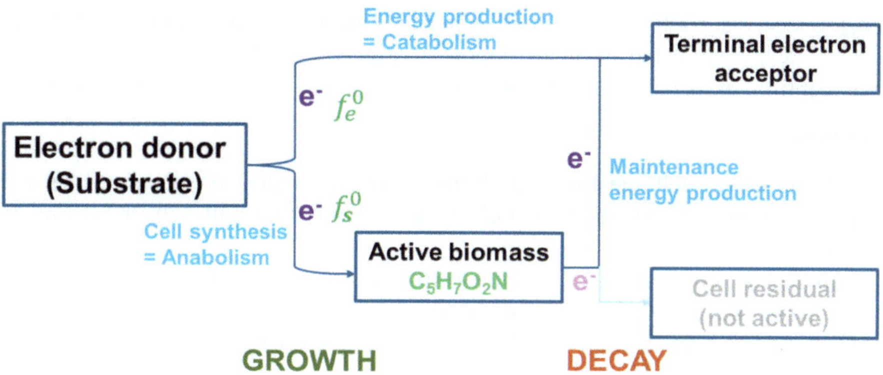

Fig. 5.1 Electron partitioning from substrate

Step 2: Balance the electrons.

Step 3: Multiply Eq. 5.1 by f_s and Eq. 5.2 by f_e

Step 4: Add Eqs. (5.1), (5.2) and (5.3) $\left[R = f_eR_a + f_sR_c - R_d\right]$

Electron partitioning can be represented with catabolism and anabolism as shown in Fig. 5.1.

Anabolism is a foundational pillar of microbial activity in biological wastewater treatment, working synergistically with catabolic reactions to facilitate both energy conservation and biomass generation. While catabolism supplies the energy and reducing power necessary for life, it is through anabolism that microorganisms build the cellular machinery to maintain their population, enzymatic activity, and structural integrity. Accurate modeling of anabolic processes is indispensable for designing and operating efficient treatment systems, especially as regulatory and environmental pressures demand more sustainable and resource-conscious approaches. The ability to write, interpret, and apply stoichiometric equations that integrate both catabolism and anabolism lies at the heart of modern environmental biotechnology. As tools such as metabolic flux analysis and systems biology become more integrated into process modeling, a deep understanding of anabolic pathways will further enhance our ability to engineer microbial communities tailored for optimal performance (Vanrolleghem et al. 1999; Henze et al. 2008). Mastering anabolic principles, therefore, not only advances academic understanding but also underpins the innovation and scalability of biological treatment processes in practice.

Practice:

Ammonium ion (NH_4^+) is released by fish and becomes toxic at low concentrations. Therefore, in order to grow fish in a closed pond, ammonium must be removed from the water. This is normally done by employing nitrifying bacteria that oxidize the ammonium to nitrate (NO_3^-), while reducing dissolved oxygen to water (nitrification). Write the overall metabolic reaction assuming$f_s = 0.1$. Let's assume this nitrification reactor receives 1000 m^3/d of pond water at ammonium concentration of 23 mg NH_4^+ − N/L, while the effluents contain 1 mg NH_4^+ − N/L. Calculate: oxygen consumption, mass of produced cells and nitrate–N concentration in the effluents.

Solution:

In order to solve such questions, firstly its important to identify substrate (that will help in writing $-R_d$ equation) and electron acceptor (that will help in writing R_a equation).

Substrate is ammonium ion, while electron acceptor is O_2 since is it nitrification where dissolved oxygen is getting reduced to form water.

Cell synthesis reaction (R_c) will be one forming $C_5H_7O_2N$ with substrate NH_4^+ and CO_2.

$$R = f_eR_a + f_sR_c - R_d$$

Since $f_s = 0.1, f_e = 0.9$

Let's write these three reactions.

$$R_c : NH_4^+ + 5CO_2 + 19H^+ + 20e^- \rightarrow C_5H_7O_2N + 8H_2O$$

$$R_a : O_2 + 4H^+ + 4e^- 2H_2O$$

$$-R_d : NH_4^+ + 3H_2O \rightarrow NO_3^- + 10H^+ + 10e^-$$

From above equations:

While adding f_sR_c, f_eR_e and $-R_d$, the first aim should be to cancel all electrons, thus need to multiple with appropriate numbers.

$$f_sR_c : NH_4^+ + 5CO_2 + 19H^+ + 20e^- \rightarrow C_5H_7O_2N + 8H_2O \times 0.1 \times 2$$

$$f_eR_a : \; O_2 + 4H^+ + 4e^- \rightarrow 2H_2O \times 0.9 \times 10$$

$$-R_d : NH_4^+ + 3H_2O \rightarrow NO_3^- + 10H^+ + 10e^- \times 4$$

Now, let's multiply and add:

$$\begin{aligned} &f_sR_c : 0.2NH_4^+ + CO_2 + 3.8H^+ + 4e^- \rightarrow 0.2C_5H_7O_2N + 1.6H_2O \\ &f_eR_a : 9O_2 + 36H^+ + 36e^- \rightarrow 18H_2O \\ &-R_d : 4NH_4^+ + 12H_2O \rightarrow 4NO_3^- + 40H^+ + 40e^- \\ \hline &R : 4.2NH_4^+ + CO_2 + 9O_2 \rightarrow 0.2C_5H_7O_2N + 4NO_3^- + 0.2H^+ + 7.6H_2O \end{aligned}$$

From overall metabolic equation, it is clear that, for each 4.2 (14) = 58.8 g ammonium ions, 9 (32) = 288 g O_2 is consumed.

Also, 0.2 (113) = 22.6 g cells and 4 (14) = 56 g $NO_3^- - N$ are produced.

From given data in the question, change in ammonium concentration = $23NH_4^+ - N/L - 1NH_4^+ - N/L = 22NH_4^+ - N/L$.

Thus,

$$\begin{aligned}\text{Amount of } NH_4^+ - N\text{treated} &= 22\,\text{mg/L} \times 1000\,\text{m}^3 \times 10^3\,\text{L/m}^3 \times \text{Kg}/10^6\,\text{mg} \\ &= 22\,\text{Kg}\end{aligned}$$

$$O_2\ \text{consumption} = 22\ \text{Kg} \times (288\,\text{g}/58.8\,\text{g}) = 107.76\,\text{Kg}$$

$$\text{Cell dry weight produced} = 22\ \text{Kg} \times (22.6\,\text{g}/58.8\,\text{g}) = 8.46\,\text{Kg}$$

$$\text{Effluent } NO_3^- - N\ \text{concentration} = 22\ \text{mg/L} \times (56\,\text{g}/58.8\,\text{g}) = 20.95\,\text{mg/L}$$

Practice:

In an aerobic activated-sludge reactor treating domestic wastewater, the organic substrate can be represented by the empirical formula $C_{10}H_{19}O_3N$ and the biomass by $C_5H_7O_2N$.

Assuming an electron-partitioning ratio of $f_e : f_s = 0.54 : 1$ (where f_e is the fraction of electrons driving respiration and f_s is the fraction driving biomass synthesis):

1. Determine how many electrons are allocated to each half-reaction.
2. Scale each half-reaction accordingly.
3. Combine them to write the overall balanced metabolic reaction, showing coefficients for O_2, NH_3, CO_2, biomass ($C_5H_7O_2N$), and H_2O.
4. Calculate the biomass yield Y in g biomass per g substrate (use molar masses of 201 g/mol for $C_{10}H_{19}O_3N$ and 113 g/mol for $C_5H_7O_2N$).

Solution:

First step: Write three half reactions

Substrate oxidation (catabolism):$-R_d$

$$C_{10}H_{19}O_3N + 17H_2O \rightarrow 10CO_2 + NH_3 + 50e^- + 50H^+ \qquad (R1)$$

Biomass synthesis (anabolism): R_c

$$NH_3 + 5CO_2 + 20H^+ + 20e^- \rightarrow C_5H_7O_2N + 8H_2O \qquad (R2)$$

Energy-yielding respiration: R_a

$$O_2 + 4H^+ + 4e^- \rightarrow 2H_2O \qquad (R3)$$

Step 2: Electron partitioning

Total electrons released by (R1): $50\,e^-$.

Let the fraction used for anabolism be $f_s = 1$unit, and for energy be $f_e = 0.54$ units.

Hence the total "electron-units" $= 1 + 0.54 = 1.54$.

Each unit corresponds to $50/1.54 = 32.47e^-$.

- Electrons for anabolism: 1 unit $\times\ 32.47e^- = 32.47e^-$
- Electrons for energy: 0.54 unit $\times\ 32.47e^- = 17.53e^-$

To match the stoichiometry of (R2) and (R3), to write f_sR_c

- Multiply anabolic half-reaction (R2) by

$$f_{smol} = 32.47/20 = 1.62$$

- Multiply energy half-reaction (R3) by

$$f_{emol} = 17.53/4 = 4.38$$

Step 3: Add reactions to write complete metabolic reaction

Reaction	Scale	Reactants	Products
R1	1	$C_{10}H_{19}O_3N + 17H_2O$	$10CO_2 + NH_3 + 50e^- + 50H^+$
R2	1.62	$1.62NH_3 + 8.11CO_2 + 32.47H^+ + 32.47e^-$	$1.62C_5H_7O_2N + 12.98H_2O$
R3	4.38	$4.38O_2 + 17.53H^+ + 17.53e^-$	$8.76H_2O$

- Electrons (e^-) and protons (H^+) cancel perfectly (50 out, 50 in).
- NH_3 and CO_2 produced by (1) partly feed (2), leaving net NH_3 consumption and CO_2 production.

$$C_{10}H_{19}O_3N + 0.62NH_3 + 4.38O_2 \rightarrow 1.62C_5H_7O_2N + 1.89CO_2 + 4.75H_2O$$

Step 4: Biomass yield

- Molar yield: 1.62 mol biomass per 1 mol substrate

- Mass yield:

$$Y = 1.62 \times \frac{M_{\text{biomass}}}{M_{\text{substrate}}} = 1.62 \times \frac{113\,\frac{\text{g}}{\text{mol}}}{201\,\frac{\text{g}}{\text{mol}}} = 0.91\,\frac{\text{g biomass}}{\text{g substrate}}$$

Step 5: Electron and oxygen balances

- Electrons: 50 released = 32.47 to anabolism + 17.53 to energy
- O_2 demand: from (R3) scale

$$4\text{e}^{-} per \text{O}_2 \Rightarrow \frac{17.53}{4} = 4.38\,\text{mol O}_2$$

All balances close, confirming a thermodynamically consistent reaction suitable for reactor sizing and yield estimation in aerobic wastewater treatment.

Further Readings

Henze M, van Loosdrecht MCM, Ekama GA, Brdjanovic D (2008) Biological wastewater treatment: principles. IWA Publishing, Modelling and Design

Vanrolleghem PA, Spanjers H, Petersen B, Ginestet P, Takács I (1999) Estimating (combinations of) activated sludge model no. 1 parameters and components by respirometry. Water Sci Technol 39:195–214

Chapter 6
Microbial Kinetics

Learning Objectives:

1. *Understand the key phases of microbial growth and how to calculate generation time and growth rate constants.*
2. *Apply kinetic models—including the Monod equation and yield coefficients—to describe substrate utilization and biomass production.*

6.1 Introduction to Microbial Kinetics

Kinetics, deriving from the Greek *κινέω* ("to move" or "to force to move"), is the study of the rates and mechanisms of processes, whether they be physical, chemical, or biological. When applied to microbial systems, microbial kinetics examines how fast biological processes occur and the mechanisms by which microorganisms grow, adapt, and interact with their environment. This chapter focuses on the quantitative description of microbial growth—including stages such as lag, exponential, stationary, and death phases—as well as the models used to predict growth and substrate consumption in systems like wastewater treatment.

The microbial kinetics is related to growth curve that was discussed in Chap. 1 (Figure Bacterial growth curve) with four distinct growth phases: the lag phase, the exponential or log phase, the stationary phase, and the death or decline phase. The exponential or log phase is important here as microbial cells proceed into cell division, where 1 cell becomes 2 cells, becomes 4, becomes 8, and so on, and the maximum growth rate is obtained during this phase. Due to the predictability of growth in this phase, this phase can be used to mathematically calculate the time it takes for the bacterial population to double in number, known as the generation time (g). The generation time g (the time required for the population to double) can be determined from the number of generations n that occur in a particular time interval

V. Achal, *Notes on Biological Processes in Environmental Engineering*,
https://doi.org/10.1007/978-981-95-1610-0_6

t, represented by t/n, with t being the specified period of time in minutes, hours, days, or months.

Let:

N_0 the initial population number
N_t population at time t
n the number of generation in time t

Then,

$$n = \frac{\log_{10} N_t - \log_{10} N_0}{0.301}$$

$$n = 3.3(\log_{10} N_t - \log_{10} N_0)$$

And, generation time (g):

$$g = \frac{t}{n} = \frac{t}{3.3(\log_{10} N_t - \log_{10} N_0)}$$

During exponential growth, the growth rate constant (k) can further be calculated. For bacterial population growth,

$$N_t = N_0 \times 2^n$$

Solving for n, the number of generations, where all logarithms are to the base 10,

$$\log N_t = \log N_0 + n \log 2N_t$$

$$n = \frac{\log N_t - \log N_0}{log2} = \frac{\log N_t - \log N_0}{0.301}$$

The growth rate constant (k),

$$k = \frac{n}{t} = \frac{\log N_t - \log N_0}{0.301t}$$

Practice: In a culture of *E. coli*, an inoculum was taken at 10 AM that contained 1000 cells/mL. A second time at 8 PM had 10,000 cells/mL. Calculate generation time for bacterial cells.

Solution:

$$n = 3.3(\log_{10} N_t - \log_{10} N_0) = 3.3((\log_{10} 10000 - \log_{10} 1000)$$
$$= 3.3(4 - 3) = 3.3$$

Generation time (g)

$$g = \frac{t}{n} = \frac{10}{3.3} = 3.01h$$

6.2 The Role of Substrate in Microbial Growth

Microorganisms depend on an external "food" source—the substrate—to fuel all of their metabolic activities. In most biodegradation and bioconversion processes, this substrate also serves as the primary electron donor, and its concentration ultimately governs how fast the biomass can reproduce. We denote:

- S = substrate concentration (e.g., g/L)
- X = active biomass concentration (e.g., g/L)

A fundamental model that relates these quantities is based on the mass balance on biomass and substrate. The rate of microbial growth is typically linked to substrate consumption through kinetic equations, as described in the next section.

As said many times earlier, microbes need food to carry out all the metabolic processes. The food is referred to as 'substrate', denoted by 'S'. The biomass (denoted by 'X') grows after utilizing substrate (S). A microbial process model is based on the mass balance on (i) the active biomass and (ii) the primary substrate that limits the growth rate of the biomass. In the vast majority of cases, the rate limiting substrate is the e^- donor. So, the term substrate is also referred as the primary e^- donor substrate.

6.3 Monod Equation

The study of the growth of bacterial cultures does not constitute a specialised subject of branch of research: it is the basic method of Microbiology (Monod 1949).

J. Monod in 1942 refined and calibrated the method for quantifying microbial growth using turbidimetry and demonstrated how growth can be mathematically described in terms of growth yield, specific growth rate, and substrate concentration (Monod 1949; Mateles and Chian 1969; Egli et al. 1993).

In 1969, Mateles and Chian conducted continuous culture experiments to investigate whether the growth dynamics observed in pure microbial cultures could also be reproduced in natural, mixed populations. Their findings showed qualitatively similar results to those from pure cultures, though with possible shifts in community composition. Moreover, in batch cultures, heterogeneous microbial populations demonstrated sequential substrate uptake patterns typical of diauxic growth (Mateles and Chian 1969).

Building on the foundational understanding of microbial growth kinetics, Thomas Egli and colleagues later emphasized the importance of precise cultivation techniques and quantification of microbial growth parameters. Their work underscored the necessity of controlling specific growth rates and utilizing defined media to ensure reproducibility and physiological relevance, particularly in continuous culture systems. They also reaffirmed the significance of Monod's model in describing microbial growth behavior (Egli et al. 1993).

During the last half century, the concepts in microbial growth kinetics have been dominated by the relatively simple empirical model proposed by Monod. Why?

The Monod's model and its kinetic parameters are sufficient for explaining growth or degradation phenomena for the majority of applications.

In this model the specific growth rate (μ) is linked to the concentration of growth controlling substrate (S) via two parameters: the maximum specific growth rate ($\mu_{\max}$) and the substrate affinity constant (K_s).

$$\mu = \mu_{\max}\frac{S}{K_s + S}$$

μ = specific biomass growth rate (per time; per hour; per day)

The specific growth rate period is defined as the rate of increase of biomass of a cell population per unit of biomass concentration.

It can be calculated by linear regression of the ln (OD at time/initial OD) versus time for the exponential growth phase. [OD = Optical Density that is calculated at 600 nm].

$\mu_{\max}$ = maximal specific growth rate (per time; per hour; per day)

Specific growth rate ($\mu_{\max}$, , per hour) is calculated using linear regression by natural logarithm of optical density / initial optical density ($\ln OD/OD_i$) versus time during the log phase, obtaining the equation $\left[\ln \frac{OD}{OD_i} = \mu_{\max} \cdot t + a\right]$ K_s–conc. giving one-half the maximum rate.

The specific growth rate (μ) can further be expressed as below:

$$\mu = \mu_{\max}\frac{S}{K_s + S} = \frac{dX}{Xdt}$$

True yield coefficient (Y):

If we assume there is no decay, the true yield (Y) can be expressed as:

$$Y = -\frac{\mathrm{d}X}{\mathrm{d}S}$$

$$\mathrm{d}X = -Y\mathrm{d}S$$

$$\frac{dX}{dt} = -Y\frac{dS}{dt}$$

From specific growth rate (μ) formula:

$$\mu = \frac{dX}{Xdt} = -Y\frac{dS}{Xdt}$$

$$-\frac{dS}{Xdt} = \frac{\mu}{Y} = q$$

where, q **= specific substrate removal rate** ($mg_S/mg_X/d$).

q is further expressed as:

$$q = q_{\max}\frac{S}{K_s + S}$$

$q_{\max}$ **= maximal specific substrate removal rate** ($mg_S/mg_X/d$)

Now, if we express biomass growth rate with specific growth rate, we can say:

$$\frac{dX}{dt} = \mu_{\max}\frac{S}{K_s + S}X = \mu X$$

Similarly, substrate utilization rate:

$$rut = \frac{dS}{dt} = -q_{\max}\frac{S}{K_s + S}X = qX$$

6.4 Kinetic and Stoichiometric Parameters Values

The kinetic parameters depend upon biomass growth and substrate utilization that have specific units and range of values. As discussed earlier, there are various types of microorganisms such as aerobic, denitrifiers, methanogens and fermenting microorganisms among others. The true yield (Y) of all these microorganisms depend on electron donor, electron acceptor, and carbon source in addition to f_s^0 (portion of electrons of an e⁻ donor substrate for cell synthesis). Table 6.1 presents typical values for various microorganisms, where the true yield (Y) is calculated based on a cellular volatile suspended solids (VSS_a) composition of $C_5H_7O_2N$, using NH_4^+ as the nitrogen source—except in cases where NO_3^- serves as the electron acceptor, in which case NO_3^- is also considered the nitrogen source. For example, aerobic heterotrophs have a yield of approximately 0.49 gVSS/gBOD and a maximum growth rate of 13.21/d in carbohydrate containing wastewater, with K_S and K_d values of 1 mgBOD/L and 0.151/d, respectively (Rittmann and McCarty 2001).

Table 6.1 Thermodynamic data for selected microbial metabolisms at standard conditions

Bacterial group	Electron donor	Acceptor	$Y\,(gVSS/gsubstrate)$	$\widehat{q}\ (gsubstrate/gVSS.d)$	$\widehat{\mu}\ (1/d)$	$K_S\,(mgsubstrate/L)$	$K_d\,(1/d)$
Aerobic Heterotrophs	Carbohydrate BOD	O_2	0.49	27	13.2	1	0.15
Denitrifiers	H_2	NO_3^-	0.81	1.25	1.0	1	0.15
Nitrifiers	NH_4^+	O_2	0.34	2.7	0.92	1	0.11
Methanogens	H_2	CO_2	0.45	1.1	0.5	–	–
Sulfate reducers	Acetate BOD	SO_4^{2-}	0.057	8.7	0.50	10	0.05

Adapted from Rittmann and McCarty (2001) and other microbiological sources

In many practical problems—especially in educational exercises and preliminary CSTR or batch-culture calculations—the tabulated specific rates $\hat{q}$ (read as q hat, specific electron-equivalent uptake rate) and $\hat{\mu}$ (read as μ hat, observed specific growth rate) are directly adopted as the kinetic maxima $q_{\max}$ and $\mu_{\max}$, respectively. This approximation rests on the assumption that the system is substrate-saturated ($S \gg K_S$), so that Monod-type expressions collapse to their plateau values ($\mu(S) \approx \mu_{\max}$, $q(S) \approx q_{\max}$).

By equating: $\mu_{\max} = \hat{\mu}$ and $q_{\max} = \hat{q}$ one bypasses detailed determination of half-saturation constants and yield conversions, greatly simplifying mass-balance calculations for biomass and substrate.

$\hat{q}$ is controlled largely by e^- flow to the electron acceptor, termed as $\hat{q}_e$.

$$\hat{q} = \frac{\hat{q}_e}{f_e^0}$$

Typical $\hat{q}$ values are presented in Table 6.1, when $\hat{q} = 1e^{-}eq/gVSS.d$

Let's take value of aerobic heterotrophs with electron donor 'carbohydrate BOD' and O_2 as electron acceptor. f_s^0 here is 0.7, means $f_e^0 = 0.3$.

$$\hat{q} = \frac{\hat{q}_e}{f_e^0} = \frac{8\frac{gBOD}{gVSS \cdot d}}{0.3} = 27\frac{gBOD}{gVSS \cdot d}$$

Why $\hat{q}_e$ is 8 in above calculation? Since one equivalent of oxygen is 8 g of O_2, one equivalent of any electron donor is equivalent to an oxygen demand of 8 g as O_2.

Table 6.2 compares the true kinetic parameters ($\hat{\mu}$ and$\hat{q}$) with their corresponding maximum values ($\mu_{\max}$ and $q_{\max}$) to highlight their conceptual and practical differences in microbial growth modeling. While $\mu_{\max}$ and $q_{\max}$ represent theoretical upper limits determined under ideal, nutrient-rich conditions (such as in batch cultures), the **true specific growth rate** ($\hat{\mu}$) and **true specific substrate utilization rate** ($\hat{q}$) reflect actual, steady-state values observed in real-world systems such as continuous-flow bioreactors or wastewater treatment processes. These true values are more relevant for engineering design and modeling, particularly in environmental biotechnology applications (Rittmann and McCarty 2001). Accurately distinguishing between these parameters is essential for developing reliable predictions of microbial performance under operational constraints.

6.4.1 Calculation of Net Growth Yield (Y_{net})

The volatile suspended solids (VSS) are derived by measuring volatile solids. Because bacteria are mostly organic, the VSS is a better indicator of organic-solids concentrations and therefore, the amount of bacteria in a sample.

Table 6.2 Comparison of true and maximum microbial kinetic parameters: $\hat{\mu} vs \cdot \mu_{\max}$ and $\hat{q} vs \cdot q_{\max}$

Parameter	Symbol	Units	Definition	Condition	Used in
True growth rate	$\hat{\mu}$	d^{-1}	Actual growth rate under steady-state	Limited energy	Process modeling
Maximum growth rate	$\mu_{\max}$	d^{-1}	Max rate in ideal/ batch culture	Abundant substrate	Lab experiments
True substrate utilization rate	$\hat{q}$	$gS/gX \cdot d$	Real substrate uptake under process conditions	Steady-state	WWTP models
Maximum substrate utilization rate	$q_{\max}$	$gS/gX \cdot d$	Max uptake when substrate is unlimited	Ideal batch growth	Physiological studies

The true yield (Y) is obtained when there is no decay (no bacterial cell death or growth phase not entering into death/decay phase). Thus, Y is also expressed as:

$$Y = -\frac{dX}{dS}$$

$$dX = -YdS$$

From here, bacterial growth rate can be expressed as:

$$\frac{dX}{dt} = -Y\frac{dS}{dt}$$

However, practically it's not possible to achieve true yield as growth curve will enter into decay or death phase under most of conditions. Under such conditions, the term 'net growth yield (Y_{net})' and 'net growth rate (μ_{net})' are used.

$$\left(\frac{dX}{dt}\right)_{net} = \left(\frac{dX}{dt}\right)_T - K_d X = -Y_T\left(\frac{dS}{dt}\right) - K_d X$$

$$\left(\frac{dX}{Xdt}\right)_{net} = \left(X\frac{dX}{dt}\right)_T - K_d$$

Thus, $\mu_{net} = \mu_T - K_d$.
Where, K_d= specific decay coefficient $(1/d)$.
Further,

$$\left(\frac{dX}{dt}\right)_{net} = -Y_T\left(\frac{dS}{dt}\right) - K_d X \quad \text{divide by} \quad -\left(\frac{dS}{dt}\right)$$

$$\frac{\left(\frac{dX}{dt}\right)_{net}}{\left(-\frac{dS}{dt}\right)} = -\frac{Y_T\left(\frac{dS}{dt}\right)}{\left(-\frac{dS}{dt}\right)} - \frac{K_d X}{\left(-\frac{dS}{dt}\right)}$$

$$-\frac{dX_{net}}{dS} = Y_T - \frac{K_d X}{\left(-\frac{dS}{dt}\right)} = Y_T - \frac{K_d}{q}$$

From the above equations:

$$Y_{net} = -\frac{\mathrm{d}X_{net}}{\mathrm{d}S} = Y_T - \frac{K_d}{q} = \frac{\mu_T}{q} - \frac{K_d}{q} = \frac{\mu_{net}}{q}$$

6.4.2 *Effect of Temperature on Biomass Growth Rate and Substrate Utilization Rate*

The kinetics of a process are related to temperature through the Arrhenius relationship (Fogler 2005). Microorganisms have developed molecular traits to respond to changing environmental temperatures. Temperature affects $\mu_{\max}$ and $q_{\max}$ that can be expressed mathematically with the Arrhenius equation by simply replacing the rate constant k in equation (with the growth rate or substrate utilization rate.

Arrhenius equation:

$$k = A\mathrm{e}^{-\frac{E_a}{RT}}$$

where k is the rate constant, s^{-1}, for a first-order rate constant; A is called a pre-exponential frequency or collision factor, s^{-1}, for a first-order rate constant; E_a is an empirical parameter, the (Arrhenius) activation energy, J mol^{-1}, characterizing the exponential temperature dependence of k; R is the universal gas constant ~8.314, J mol^{-1} K^{-1}; and T is the absolute temperature, K.

Further,

$$\ln k = \frac{-E_a}{RT} + \ln A$$

The above equation can further be expressed in bacterial growth when temperature changes from one ($T1$) to another ($T2$).

Accordingly, $k1$ at $T1$:

$$\ln k1 = \ln A - \frac{E_a}{RT1}$$

$k2$ at $T2$:

$$\ln k2 = \ln A - \frac{E_a}{RT2}$$

Integration between $T1$ and $T2$ gives:

$$\ln k_2 - \ln k_1 = \frac{-E_a}{R}\left[\frac{1}{T_2} - \frac{1}{T_1}\right]$$

$$\ln \frac{k_2}{k_1} = \frac{-E_a}{R}\left[\frac{1}{T_2} - \frac{1}{T_1}\right]$$

$$\ln \frac{k_2}{k_1} = const(T_2 - T_1)$$

$$\frac{k_2}{k_1} = \mathrm{e}^{const(T_2 - T_1)}$$

$$e^{const} \textit{ is denoted by } \theta$$

$$\frac{k_2}{k_1} = \theta^{(T_2 - T_1)}$$

Practice: If a reaction has an activation energy of 50 kJ/mol, then how much should the rate of the reaction accelerate if the temperature is raised from 300 to 310 K?

Solution:

$$\ln\left(\frac{K_{310}}{K_{300}}\right) = \frac{-5000\,\mathrm{J}}{8.314\frac{\mathrm{J}}{\mathrm{mol.K}}}\left(\frac{1}{310} - \frac{1}{300}\right)K = 0.647$$

$$K_{310} = e^{0.647} K_{300} = 1.9K_{300} \approx 2K_{300}$$

The rate of reaction would double when temperature is raised from 300 to 310 K.

Since the rate constant is now a function of the temperature difference, it is no longer necessary to convert the temperature to Kelvin, and, if 20 °C is the reference temperature, it will yield (Sheridan et al. 2012):

$$k = k_{20} \cdot \theta^{(T-20)}$$

The above equation is also applied to calculate substrate-utilization rate or biomass growth rate at different temperatures. For temperatures up to the bacterial optimum temperature, the substrate-utilization rate or biomass growth rate roughly doubles for each 10 °C increase in the temperature. This phenomenon can be approximated by

$$\mu_{\max}(T) = \mu_{\max}(20\,°\mathrm{C}) \cdot \theta^{(T-20)}$$

$$q_{\max}(T) = q_{\max}(20\,°\mathrm{C}) \cdot \theta^{(T-20)}$$

where T is in °C and $\mu_{max}(20\,°C)$ or $q_{max}(20\,°C)$ is the maximal value for 20 °C.

The value of θ for 'activated sludge' is typically 1.047.

The similar formula is also applicable to calculate BOD at changing temperature that can be understood better with practice question below.

Practice

To test the biochemical oxygen demand (BOD), A 10 mL sample of wastewater was taken. In the beginning of the BOD experiment the dissolved oxygen concentration was 9 mg/L, after 5 days it decreased to 3 mg/L.

The volume of the test bottle is equal to 300 ml and the reaction constant at 20 ℃ is.

$k = 0.22\ \text{day}^{-1}$. Calculate BOD_5 at 30° C.

Solution

The BOD rate constant (k) is adjusted to the temperature of receiving water using following relationship:

$$k_{T2} = k_{T1} \cdot 1 \cdot 047^{(T2-T1)}$$

Initial dissolved oxygen = 9 mg/L
Final dissolved oxygen = 3 mg/L
Time (t) = 5 days.
Reaction constant (k) at 20 °C = 0.22 day^{-1}
Now, calculate the oxygen demand exerted initially (L_0).

$$L_0 = 9 - 3 = 6\,\text{mgO}_2/\text{L}$$

Now, adjust the reaction constant for 30 °C.

$$k_{30\,°C} = 0.22[1/\text{day}] * 1.047^{30\,°C-20\,°C} = 0.35[1/\text{day}]$$

Now BOD_5 at 30°**C** can be calculated as:

$$BOD_5 = 6 \times (1 - e^{-0.35\times5)} = 4.95\,\text{mgO}_2/\text{L}$$

Practice

Assume, a reactor containing 200 mg/L of biomass consumes 4 mg/L acetate in 10 min. If the true yield of the culture in this reactor is 0.5 g cells per g acetate, and the decay constant is 0.1 1/d; determine:

(i) Substrate utilization rate, *rut*, in mg Acetate/(L.d).
(ii) Specific substrate utilization rate.
(iii) True and net specific biomass growth rate (μ_T, μ_{net}).
(iv) Net yield (Y_{net}).
(v) Daily amount of sludge to be removed to maintain constant X.

Solution

Given:

- Biomass concentration, $X = 200\,\text{mg/L}$.
- Acetate consumption, $dS = -4\,\text{mg/L}$ (minus symbol because substrate is getting consumed) in ($t = 10$) minutes.
- True yield, $Y_T = 0.5\,\text{g cells/g acetate} = 0.5\,\text{g cells/g acetate}$.
- Decay constant, $K_d = 0.11/\text{day}$.

To solve this problem, we'll go through each part step by step, using the given data and relevant formulas.

(i)

$$rut = \frac{dS}{dt} = \frac{-4\,\text{mg/L}}{10\,\text{min}} \times \frac{60\,\text{min}}{\text{h}} \times \frac{24\,\text{h}}{\text{d}} = -576\frac{\text{mgAcetate}}{\text{L.d}}$$

(ii)

$$q = -\frac{\text{d}S}{\text{d}t.X} = \frac{-\text{d}S}{X\,\text{d}t} = \frac{576\,\text{mg Acetate/L.d}}{200\frac{\text{mg cells}}{\text{L}}} = 2.88\frac{\text{mg Acetate}}{\text{mg cells.d}}$$

(iii)

$$\mu_T = qY = 2.88 \times 0.5 = 1.44\frac{1}{d}$$

$$\mu_{net} = \mu_T - K_d = 1.44 - 0.1 = 1.34\frac{1}{d}$$

(iv)

$$Y_{net} = Y_T - \frac{Kd}{q} = 0.5 - \frac{0.1}{2.88} = 0.465\frac{\text{g cells}}{\text{g Acetate}}$$

Biomass is growing at 134% every day.
Thus, $X_{1d} = 2.34X_0 = 2.34 \times 200\,\text{mg}\frac{\text{cells}}{\text{L}} = 468\,\text{mg}\frac{\text{cells}}{\text{L}}$.
The daily amount of sludge to be removed $= 468\,\text{mg}\frac{\text{cells}}{\text{L}} - 4200\,\text{mg}\frac{\text{cells}}{\text{L}} = 268\,\text{mg}\frac{\text{cells}}{\text{L}}$.

Practice:

For production of penicillin ($C_{16}H_{18}O_4N_2S$) using *Penicillium* mould, glucose ($C_6H_{12}O_6$) is used as substrate, and phenylacetic acid ($C_8H_8O_2$) is added as precursor. The stoichiometry for overall synthesis is:

$$1.67C_6H_{12}O_6 + 2NH_3 + 0.5O_2 + H_2SO_4 + C_8H_8O_2 \\ \rightarrow C_{16}H_{18}O_4N_2S + 2CO_2 + 9H_2O$$

(a) What is the maximum theoretical yield of penicillin from glucose?
(b) When results from a particular penicillin fermentation were analysed, it was found that 24% of the glucose had been used for growth, 70% for cell maintenance activities (such as membrane transport and macromolecule turnover), and only 6% for penicillin synthesis. Calculate the yield of penicillin from glucose under these conditions.
(c) Batch fermentation under the conditions described in (b) is carried out in a 100 L tank. Initially, the tank is filled with nutrient medium containing 50 g/L glucose and 4 g/L phenylacetic acid. If the reaction is stopped when the glucose concentration is 5.5 g/L, determine: (i) which is the limiting substrate if NH_3, O_2 and H_2SO_4 are provided in excess; (ii) the total mass of glucose used for growth; (iii) the amount of penicillin produced; and (iv) the final concentration of phenylacetic acid.

Solution:

Given reaction:

$$1.67C_6H_{12}O_6 + 2NH_3 + 0.5O_2 + H_2SO_4 + C_8H_8O_2 \\ \rightarrow C_{16}H_{18}O_4N_2S + 2CO_2 + 9H_2O$$

Molecular weights:

- Glucose ($C_6H_{12}O_6$): 180 g/mol
- Penicillin ($C_{16}H_{18}O_4N_2S$): 334 g/mol
- Phenylacetic acid ($C_8H_8O_2$): 136 g/mol

Step 1: Theoretical yield of penicillin from glucose

- Stoichiometry: 1 mol penicillin is made from 1.67 mol glucose
- **Molar yield**:

$$Y_{\frac{P}{S}}^{\text{mol}} = \frac{1\ \text{mol Penicillin}}{1.67\ \text{mol Glucose}} = 0.5988\frac{\text{mol Penicillin}}{\text{mol Glucose}}$$

- **Mass yield**:

$$Y_{\frac{P}{S}}^{mass} = Y_{\frac{P}{S}}^{mol} \times \frac{334\frac{g}{mol}}{180\frac{g}{mol}} = 0.5988 \times 1.8556 = 1.11\frac{\text{g Penicillin}}{\text{mol Glucose}}$$

Answer (a):$Y_{\frac{P}{S}}^{max} = 0.599\frac{mol}{mol}$ or 1.11 g Penicillin per g Glucose$Y_{\frac{P}{S}}^{max} = 0.599\frac{mol}{mol}$ or 1.11 g Penicillin per g Glucose

Step 2: Actual yield when 6% of glucose goes to penicillin

Only 6% of the consumed glucose is channeled into penicillin synthesis. Therefore:

$$Y_{\frac{P}{S}}^{actual} = 1.11\frac{\text{g Penicillin}}{\text{g Glucose}} \times 0.06 = 0.0667\frac{\text{g Penicillin}}{\text{g Glucose}}$$

Answer (b):$Y_{\frac{P}{S}}^{actual} = 0.0667\frac{\text{g Penicillin}}{\text{g Glucose}}$ or, 66.7 mg Penicillin per g Glucose$Y_{\frac{P}{S}}^{actual} = 0.0667\frac{\text{g Penicillin}}{\text{g Glucose}}$ or, 66.7 mg Penicillin per g Glucose

Step 3: Batch fermentation in 100 L tank

- Initial:
  - Glucose:50 g/L $\Rightarrow 50 \times 100 = 5000$ g total
  - Phenylacetic acid:4 g/L $\Rightarrow 4 \times 100 = 400$ g total
- Final (stop): Glucose 5.5 g/L $\Rightarrow 5.5 \times 100 = 550$ g remaining
- Glucose consumed: $5000 - 550 = 4450$ g

(i) **Limiting substrate**

For penicillin synthesis, stoichiometric ratio is 1.67 mol Glucose: 1 mol precursor

$$\frac{m_{Glucose}}{m_{Precursor}} = 1.67 \times \frac{180}{136} = 2.21\frac{\text{g Glucose}}{\text{g Precursor}}$$

Actual initial ratio $5000/400 = 12.5$, which is > 2.21, so phenylacetic acid is the limiting substrate for penicillin formation (Answer).

(ii) **Mass of glucose used for growth**

$$24\% \text{ of the } 4450 \text{ g consumed} : 0.24 \times 4450 = 1068 \text{ g (Answer)}$$

(iii) **Penicillin produced**

- Glucose to penicillin:6% of 4450 g used $\rightarrow 0.06 \times 4450 = 267$ g glucose

- Mass of penicillin from that glucose:

$$267\,\text{g Glucose} \times 1.11\,\text{g Penicillin/gGlucose} = 296.4\,\text{g Penicillin (Answer)}$$

(iv) **Final concentration of phenylacetic acid**

- Moles of penicillin made:296.4 g/334 g/mol = 0.888 mol
- Precursor consumed (1 mol precursor/mol penicillin):

$$0.888\,\text{mol} \times 136\,\text{g/mol} = 120.8\,\text{g}$$

- Remaining precursor: 400 − 120.8 = 279.2 g in 100 L ⇒ 2.79 g/L **(Answer)**

6.4.3 Factors that Affect Reaction Rate

μ_{max} and q_{max} are kinetic rates. The rate of a reaction depends on environmental conditions, such as temperature and pH.

Effect of temperature on kinetic constants (0 °C < T < 32°)

$$\mu_{max(T)} = \mu_{max(T1)}\theta^{T-T1}$$

$$1 < \theta < 1.8$$

Typically, $\theta_{\text{Activated sludge}} = 1.047.\theta_{\text{Activated sludge}} = 1.047.$

Effect of pH on kinetic constants (150 < K_{pH} < 200)

$$\mu_{max(pH)} = \mu_{max(optimalpH)}\frac{k_{pH}}{k_{pH} + I}$$

where,$I = 10^{[optimalpH - pH]} - 1.$

6.4.4 Effects of Oxygen and Nutrient Limitation on Microbial Growth Rates

In most natural and engineered environments, microorganisms rarely experience unlimited supplies of all growth-supporting factors. In addition to their carbon and energy source, they require oxygen (for aerobes) and a suite of macro- and micronutrients (e.g., nitrogen, phosphorus, sulfur, trace metals) in the correct proportions.

When any one of these becomes limiting, the observed specific growth rate μ falls below its theoretical maximum μ_{max}. Below we explore in detail how oxygen and nutrient limitations alter microbial kinetics, how those limitations can be modeled, and how they interplay in multi-limiting environments.

Oxygen Limitation

- **Monod-type Dependence on Dissolved Oxygen**

Aerobic microorganisms consume dissolved oxygen (DO) as their terminal electron acceptor. Analogous to substrate limitation, their specific growth rate under oxygen limitation can be described by a Monod expression:

$$\mu = \mu \max \frac{[O_2]}{Ko_2 + [O_2]}$$

where:

$[O_2]$ is the dissolved oxygen concentration,
K_{O_2} is the half-saturation constant for oxygen, often in the range 0.1–1 mg/L for many bacteria.

- **Mass-Transfer Constraints**

In stirred-tank reactors or large aeration basins, the rate at which oxygen dissolves into the liquid (quantified by $K_L a$ and the driving force $\Delta C = C * -[O_2]$) can be the true bottleneck. Even if $[O_2]$ in the bulk is above K_{O_2}, poor mixing or low $K_L a$ can create microzones of starvation at the cell boundary, effectively lowering the apparent μ_{max}.

- **Threshold and Critical DO**

Many aerobes exhibit a critical DO concentration below which growth essentially ceases. This threshold is sometimes modeled by combining Monod kinetics with an inhibition term:

$$\mu = \mu \max \frac{[O_2]}{K_{O_2} + [O_2] + \frac{[O_2]^2}{K_I}}$$

where K_I represents the onset of oxygen toxicity at high concentrations, though in practice most dissolved-oxygen issues are about low DO rather than inhibition by excess.

Macro-nutrient Limitation (Nitrogen, Phosphorus, Sulfur)

1. **Liebig's Law of the Minimum**

Often the single most limiting nutrient (e.g., NH_4^+-N or $PO_4^{3-}-P$) controls growth. In its simplest form, the specific growth rate can be modeled as:

$$\mu = \mu_{\max}^{\min}\left\{\frac{[C]}{K_C + [C]}, \frac{[N]}{K_N + [N]}, \frac{[P]}{K_P + [P]}\right\}$$

where each term is a Monod expression for carbon, nitrogen, or phosphorus.

2. **Multiplicative (Interactive) Model**

When two (or more) nutrients co-limit, a multiplicative form may better capture their joint effect:

$$\mu = \mu \max\left[\frac{[C]}{K_C + [C]} \cdot \frac{[N]}{K_N + [N]} \cdot \frac{[P]}{K_P + [P]}\right]$$

This model predicts that even moderate depletion in any one nutrient proportionately reduces μ.

Practice: If the maximal specific growth rate of an activated sludge culture is 0.9 1/d at 20 °C and pH = 7, what would be the maximal specific growth rate at 25 °C and pH = 6? (Given, $K_{pH} = 175, \theta = 1.047$)

Solution:

$$\begin{aligned}\mu_{\max(25)} &= \mu_{\max(20)} \cdot \theta^{(25-20)} \\ &= 0.9\frac{1}{d} \times 1.047^5 \\ &= 1.13\frac{1}{d}\end{aligned}$$

From, $I = 10^{[optimalpH - pH]} - 1$

$$I = 10^{[7-6]} - 1 = 9$$

$$\begin{aligned}\mu_{\max} &= \mu_{\max(pH)} \cdot \frac{K_{pH}}{K_{pH} + I} \\ &= 1.13\frac{1}{d} \cdot \frac{175}{175 + 9} \\ &= 1.071/d\end{aligned}$$

Further Readings

Egli T, Lendenmann U, Snozzi M (1993) Kinetics of microbial growth with mixtures of carbon sources. Antonie Van Leeuwenhoek 63(3–4):289–298

Fogler HS (2005) Elements of chemical reaction engineering. Prentice Hall, USA

Mateles RI, Chian SK (1969) Kinetics of substrate uptake in pure and mixed culture. Environ Sci Technol 3:569–574
Monod J (1949) The growth of bacterial cultures. Ann Rev Microbiol 3:371–394
Rittmann BE, McCarty PL (2001) Environmental biotechnology: principles and applications. McGraw-Hill Education, New York, Chicago, San Francisco, Athens, London, Madrid, Mexico City, Milan, New Delhi, Singapore, Sydney, Toronto
Sheridan C, Jochen Petersen J, Rohwer J (2012) On modifying the Arrhenius equation to compensate for temperature changes for reactions within biological systems. Water SA 38:149–151

Chapter 7
Bioreactor

Learning Objectives:

1. *Describe the design, classification, and operational modes of bioreactors, and explain how they create an optimal environment for microbial processes.*
2. *Apply mass balance principles and kinetic equations to analyze reactor performance, including calculations for hydraulic retention time, substrate utilization, and the food-to-microorganism ratio (F/M).*

7.1 Introduction to Bioreactors

A bioreactor is a meticulously engineered system designed to support the growth of micro organisms or cells under carefully controlled conditions to facilitate specific biochemical reactions. Acting as the core of biotechnological processes, bioreactors enable the conversion of substrates into valuable products such as pharmaceuticals, enzymes, biofuels, organic acids, polymers, and other metabolites. These reactors ensure optimal and reproducible conditions that enhance productivity, product yield, and process efficiency.

Typical bioreactors are cylindrical vessels, equipped with mechanisms for agitation, aeration, temperature regulation, pH control, nutrient supply, and waste removal. They vary significantly in size, ranging from laboratory-scale reactors holding less than one liter, to massive industrial reactors exceeding 50,000 L in volume. The materials used for bioreactor construction—ranging from stainless steel, glass, glass-lined steel, to polymeric materials—depend on process specifications, sterility requirements, chemical resistance, and scale.

Bioreactors must precisely regulate several environmental conditions critical to microbial or cellular performance, including:

- Temperature: maintained through heating or cooling jackets to optimize enzymatic activities and metabolic rates.

V. Achal, *Notes on Biological Processes in Environmental Engineering*,
https://doi.org/10.1007/978-981-95-1610-0_7

- pH: regulated via automated addition of acids or bases to ensure ideal conditions for microbial growth and product formation.
- Dissolved oxygen (DO): monitored and controlled through aeration and agitation mechanisms to fulfill aerobic metabolism demands.
- Nutrient concentrations: meticulously adjusted to avoid nutrient deficiency or excess, thereby preventing metabolic inhibition or unwanted by-products.
- Mixing: achieved by mechanical agitators or circulation pumps, ensuring homogeneous conditions throughout the reactor.

Key factors influencing bioreactor design and operation include:

- Microbial and Cell Type: Bacteria, yeast, fungi, algae, or mammalian cells each require specific growth environments, influencing aeration, mixing intensity, and nutrient provision strategies.
- Desired Product: High-value products such as recombinant proteins and pharmaceuticals often demand stringent sterility, while bulk chemicals may prioritize economic efficiency.
- Operational Mode: Choices among batch, continuous, fed-batch, or perfusion modes depend on productivity goals, economic considerations, product stability, and process scalability.

7.2 Classification of Bioreactors

Based on the mode of operation, bioreactors are classified into three types (Fig. 7.1):

(i) Batch operation: Under this mode, reactor is filled with nutrient medium and allowed for reactions to proceed. Bioreactor is emptied at the end, cleaned, re-filled, re-inoculated with microbial strain and start again.
(ii) Continuous operation: Under this mode, there is continuous input of nutrients and continuous output of products.
(iii) Fed-batch operation: This mode is between batch and continuous operations. Under this mode, fresh nutrient media is continuously or sometimes periodically added to bioreactor and it is emptied or partially emptied when reactor is full.

Further, bioreactors are generally classified into two broad categories:

(i) Suspended Growth Bioreactors: In these systems, microbial cells grow freely suspended in the medium. They are ideal for homogeneous mixing and precise control of environmental parameters.

   - Batch reactors: Reactants are added once, and products are harvested at the end of the reaction. Simple design but lower productivity.
   - Fed-batch reactors: Nutrients are gradually added, allowing enhanced control of substrate concentration, improved yields, and reduced inhibitory effects.

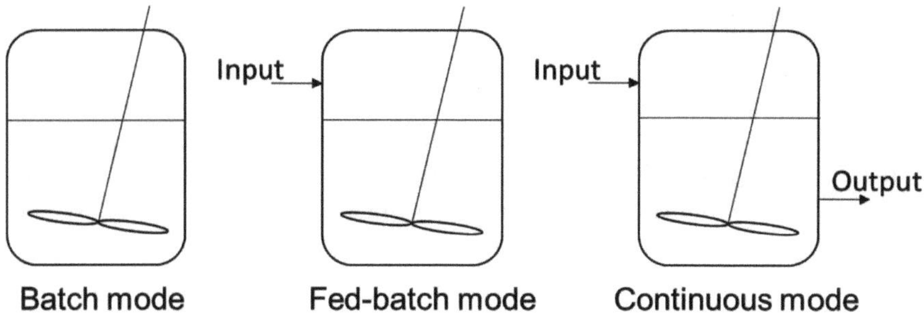

Fig. 7.1 Mode of operation of bioreactors

- Continuous Stirred Tank Reactors (CSTRs): Continuous inflow and outflow of medium maintain steady-state conditions, achieving consistent product quality and higher productivity.
- Plug Flow Reactors (PFRs): Medium flows continuously through a tubular reactor without back-mixing, providing gradients of substrate concentration and allowing reaction optimization along the reactor length.

(ii) Attached Growth or Biofilm Bioreactors: These reactors utilize microbial biofilms attached to solid surfaces or carriers, facilitating the treatment of wastewater, production of specialty chemicals, or bioremediation processes. The biofilm mode promotes microbial resilience and process stability.

- Packed Bed Reactors: Utilize fixed beds of packing material that support biofilms; effective for processes like nitrification, denitrification, or biochemical transformations.
- Fluidized Bed Reactors: Particles carrying biofilms are fluidized by upward fluid flow, enhancing mass transfer and reaction rates.
- Membrane Bioreactors (MBRs): Integrate biological processes with membrane filtration, offering high-quality effluent treatment.
- Airlift Reactors: Rely on air injection for mixing, reducing shear stress, and suitable for delicate cells or microorganisms.
- Upflow Anaerobic Sludge Blanket (UASB) Reactors: Promote granular biofilm formation under anaerobic conditions, highly efficient for wastewater treatment and biogas generation.

Advanced bioreactor designs incorporate sophisticated sensors, automated control systems, and computational modeling to optimize process conditions in real-time, significantly enhancing productivity, scalability, and reproducibility. Recent developments, including single-use disposable bioreactors, reduce sterilization and maintenance costs, minimize contamination risks, and expedite biopharmaceutical manufacturing (Jaibiba et al. 2020).

7.3 Mass Balances

The mass balance is the key to design and analysis of biological processes in bioreactor. One type of mass balance is provided by balanced chemical equation. For example, when denitrifying bacteria are employed to treat wastewater containing benzoate and 40% of the electron equivalents in benzoate is used for cell synthesis ($f_s = 0.40$), while ammonium is available as the nitrogen source, following biological reaction is obtained.

$$0.0333C_6H_5COO^- + 0.12NO_3^- + 0.02NH_4^+ + 0.12H^+ \rightarrow 0.02C_5H_7O_2N + 0.06N_2 + 0.12CO_2 + 0.0133HCO_3^- + 0.1067H_2O$$

The above reaction indicates that for each 0.033 mol of benzoate consumed by denitrifying bacteria, 0.12 mol nitrate is required to serve as an electron acceptor and 0.02 mol ammonium ion serves as a nitrogen source for bacterial growth that produces 0.02 mol of bacteria along with defined amount of nitrogen and carbon dioxide gases, bicarbonate ion, and water.

The 0.02 mol of bacteria produced in the reaction represents a waste product, termed as sludge, waste biomass, waste biosolids, or excess biosolids. This must be removed from the system and disposed properly that may require further treatment it is important to know the quantity of waste sludge to design sludge-disposal facility.

7.3.1 Mass Conservation in Reactive Flow Systems

Conservation of mass states that matter cannot be created or destroyed, only changed in form. This concept is applied to chemical reactions, which involve the transformation of substances (reactants) into different substances (products). During a chemical reaction, the total mass of the reactants is equal to the total mass of the products. This is due to the fact that atoms are conserved during a chemical reaction and cannot be created or destroyed.

The law of conservation of mass can be mathematically expressed as:

Mass of Reactants = Mass of Products.

Or, $\sum m_{reactants} = \sum m_{products}$

This statement is known as the law of conservation of mass, or sometimes referred to as Lavoisier's Law, after the French chemist Antoine Lavoisier who first proposed it in 1789. It is an important concept in chemistry as it helps to explain why chemical reactions occur in the way that they do.

For example, the reaction of hydrogen and oxygen to form water can be expressed as: $2H_2 + O_2 \rightarrow 2H_2O$.

The mass of the reactants ($2H_2 + O_2$) is equal to the mass of the products ($2H_2O$), demonstrating that mass is conserved during the reaction.

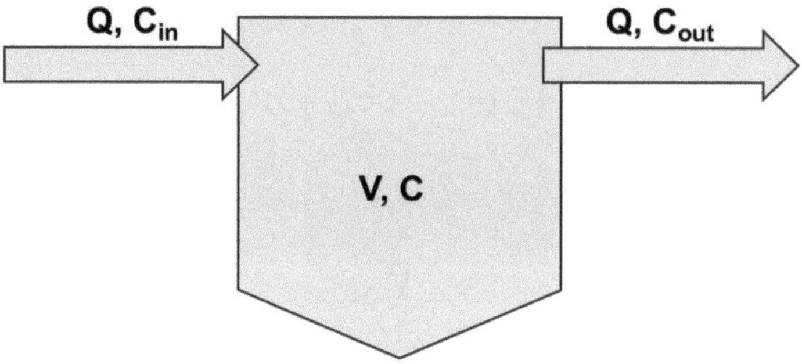

Fig. 7.2 Control volume reactor

Let's consider a perfectly mixed reactor (control volume) of constant volume V, through which a single chemical species flows and reacts. It may be represented as Fig. 7.2.

Where:

V	Volume of reactor (m^3)
C	Concentration inside reactor (kg/m^3)
Q	Volumetric flow rate in and out (m^3/s)
C_{in}, C_{out}	Inlet and outlet concentrations (kg/m^3)
r	Net reaction rate (positive if producing species) [$(kg/m^3)/s$]
dC/dt	the rate of change of the concentration of the reactants over time

For one inlet and one outlet, the general mass-baance over the reactor is:

$$\underbrace{V\frac{dC}{dt}}_{\text{Accumulation}\left(\frac{Kg}{s}\right)} = \underbrace{QC_{in} - QC_{out}}_{\text{Netflow}\left(\frac{Kg}{s}\right)} + \underbrace{rV}_{\text{Reaction}\left(\frac{Kg}{s}\right)}$$

Under steady state condition (where there is no changes in time), above formula will be changed, as d/dt = 0 that makes dC/dt = 0.

Thus:

$$0 = QC_{in} - QC_{out} + rV$$

$$rV = Q(C_{out} - C_{in})$$

$$r = \frac{Q}{V}(C_{out} - C_{in})$$

In the absence of any reaction (r = 0), it follows that $C_{in} = C_{out}$. This single-equation framework underpins the design and analysis of all continuous reactors in biochemical and chemical engineering.

Thus:

$$0 = QC_{\text{in}} - QC_{\text{out}} + rV$$

$$rV = Q(C_{\text{out}} - C_{\text{in}})$$

$$r = \frac{Q}{V}\Delta C$$

If there is no reaction, r = 0.
Then, $C_{\text{in}} = C_{\text{out}}$

7.4 Biological Batch Reactor

A biological batch reactor is a type of bioreactor commonly employed in laboratory studies and pilot-scale wastewater treatment processes to investigate microbial behavior, substrate degradation, and reaction kinetics under controlled conditions. In this reactor type, the system is **closed with respect to mass transfer**—meaning that no additional substrate or microorganisms are added or removed during the reaction phase. As such, all biological transformations take place within a fixed volume over time, making the batch reactor an ideal setup for studying time-dependent changes in substrate concentration, biomass growth, and product formation.

Mathematically, a batch reactor can be described by a first-order differential equation that relates the **substrate concentration (S)** to **time (t)**. Assuming a first-order reaction rate for substrate removal (r_{ut}), the rate equation is expressed as:

$$\frac{dS}{dt} = -kS$$

where:

S is the substrate concentration (e.g., COD, BOD, or a specific organic compound),
k is the **reaction rate constant** (day^{-1} or h^{-1}, depending on the time unit used),
$\frac{dS}{dt}$ is the **substrate utilization rate** or rate of degradation.

The integration of this equation yields an exponential decay of substrate concentration over time:

$$S_{(t)} = S_0 e^{-kt}$$

where S_0 is the initial substrate concentration.

In biological systems, however, substrate removal is rarely governed by simple first-order kinetics. Instead, microbial growth and substrate consumption typically follow **Monod kinetics**, where the rate of substrate utilization is a function of both the substrate concentration and microbial activity. The Monod equation is often used to model these systems:

$$\frac{dS}{dt} = -q\max\frac{S}{Ks + S}X$$

where:

$q\max$ is the maximum specific substrate utilization rate (g substrate/g biomass·day),
X is the biomass concentration (g/L),
K_S is the half-saturation constant (g/L), indicating the affinity of the microorganisms for the substrate.

Batch reactors provide a powerful tool for determining key kinetic parameters such as **growth rate (μ)**, **yield (Y)**, **substrate affinity** (K_S), and **decay coefficients**, which are essential for the design and optimization of full-scale biological treatment systems.

Several studies (Egli 1995; Doran 2013; Rittmann and McCarty 2001) have used batch systems to characterize microbial metabolism under different environmental conditions. Batch reactors are particularly useful for simulating shock loads, toxic inhibition, or substrate switching events (e.g., diauxic growth), making them valuable for environmental microbiology and bioprocess engineering.

7.5 Continuous Flow Reactor

A continuous flow reactor is a type of reactor where reactants are continuously added to the reactor over time, and the products and reactants are continually removed from the reactor. This type of reactor is commonly used in chemical engineering and chemical manufacturing processes.

The general equation for a continuous flow reactor is:

$$V\frac{\mathrm{d}C}{\mathrm{d}t} = QC_{\mathrm{in}} - QC_{\mathrm{out}} + rV = QC_{\mathrm{in}} - QC_{\mathrm{out}} - kCV$$

This equation can be used to determine the concentration of reactants in a continuous flow reactor as a function of time.

Under steady state condition:

$$0 = QC_{\mathrm{in}} - QC_{\mathrm{out}} - kCV$$

In continuous stirred tank reactor (CSTR): $C = C_{\mathrm{out}}$

$$0 = QC_{\mathrm{in}} - QC - kCV$$

$$QC_{\text{in}} = C(Q - kV)$$

Divide by Q to both sides

$$C = \frac{C_{\text{in}}}{1 + \frac{kV}{Q}}$$

$$C = \frac{C_{\text{in}}}{1 + k\theta}$$

where $\theta = \frac{V}{Q}$ = Hydraulic Retention Time (HRT), expressed in day.

HRT is time required for the fluent feed to spend inside the reactor.

$$HRT = \frac{V}{Q} = \frac{\text{Total volume of the reactor}}{\text{Amount of feed inside the reactor}} = \frac{\text{Volume of the medium}}{\text{Flow rate into the reactor}}$$

7.6 Application of Biological Reactor

7.6.1 *Chemostat*

A chemostat is a type of biological reactor that is used to maintain a constant culture of microorganisms in a liquid medium by controlling the flow of liquid medium into the reactor and removing a portion of the culture. It follows a continuous flow scheme at steady state. It is an important tool in research and biotechnology and is used to study how microorganisms grow and interact with each other in liquid media.

7.6.2 *Mass Balance on Substrate (S)*

If mass balance is done on the substrate in Fig. 7.3:

$$0 = QS^0 - QS + r_{ut}V$$

Thus, mass balance on the substrate:

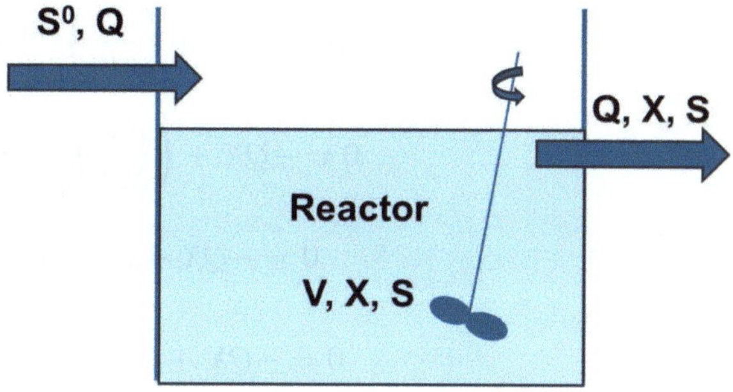

Fig. 7.3 CSTR (where S^0 = initial substrate concentration, S = substrate concentration, Q = flow, V = volume of reactor, and X = biomass concentration)

$$0 = Q(S^0 - S) - qXV$$

From microbial kinetics, we know, substrate utilization rate,

$$r_{ut} = -qX = -q_{\max} \frac{S}{K_S + S} \cdot X$$

So,

$$0 = Q(S^0 - S) - q_{\max} \frac{S}{K_S + S} \cdot XV$$

$$\frac{Q(S^0 - S)}{XV} = q_{\max} \frac{S}{K_S + S} = q$$

$$\frac{(S^0 - S)}{qX} = \theta$$

$$X = \frac{Q(S^0 - S)}{q_{\max} \frac{S}{K_S + S} V}$$

7.6.3 *Mass Balance on Biomass (X)*

If mass balance is done on the biomass in Fig. 7.3:

$$0 = QX^0 - QX + rV$$

$$X^0 = \text{Initial biomass concentration} \approx 0$$

$$0 = -QX + \left(\frac{dX}{dt}\right)_{\text{net}} \cdot V$$

$$0 = -QX + \left(\frac{dX}{Xdt}\right)_{\text{net}} \cdot XV$$

$$0 = -QX + \mu_{\text{net}} \cdot XV$$

$$0 = -QX + (\mu - K_d)XV$$

$$0 = -QX + \mu_{\max}\frac{S}{K_S + S}XV - K_dXV$$

$$\left[\text{We know} : \mu_{\max} = Yq_{\max}\right]$$

$$0 = -QX + Yq_{\max}\frac{S}{K_S + S}XV - K_dXV$$

$$0 = Yq_{\max}\frac{S}{K_S + S}XV - K_dXV - QX$$

[Divide by X]

$$0 = Yq_{\max}\frac{S}{K_S + S}V - K_dV - Q$$

[Divide by Q]

$$0 = Yq_{\max}\frac{S}{K_S + S}\frac{V}{Q} - K_d\frac{V}{Q} - 1$$

$$0 = Yq_{\max}\frac{S}{K_S + S}\theta - K_d\theta - 1$$

[Multiple by $(K_S + S)$]

$$0 = Yq_{\max}S\theta - K_d\theta(K_S + S) - K_S - S$$

$$0 = Yq_{\max}S\theta - K_d\theta K_S + -K_d\theta S - K_S - S$$

$$0 = S(Yq_{\max}\theta - K_d\theta - 1) - K_S(K_d\theta + 1)$$

$$S = \frac{K_s(K_d\theta + 1)}{Yq_{\max}\theta - K_d\theta - 1}$$

Now, equations for X can further be simplified based on the mass balance on X. If we look into the above equations, we find:

$$0 = Yq_{\max}\frac{S}{K_S + S}V - K_d V - Q$$

$$q_{\max}\frac{S}{K_S + S} = \frac{K_d V + Q}{YV}$$

From mass balance on substrate, S:

$$X = \frac{Q(S^0 - S)}{q_{\max}\frac{S}{K_S+S}V} = \frac{Q(S^0 - S)}{\left(\frac{K_d V+Q}{YV}\right).V} = \frac{QY(S^0 - S)}{\frac{Q}{Q}(K_d V + Q)} = \frac{Y(S^0 - S)}{\left(K_d\frac{V}{Q} + 1\right)}$$

$$X = \frac{Y(S^0 - S)}{(K_d\theta + 1)}$$

While HRT is a useful parameter for understanding the hydraulic dynamics of a bioreactor, it does not fully account for the biological activity of microorganisms in the system. A more critical parameter for microbial growth and treatment efficiency is the Sludge Retention Time (SRT), also known as sludge age. SRT represents the average time that activated sludge microorganism remain in the system, directly influencing biomass concentration, substrate degradation rates, and sludge settleability. Unlike HRT, which depends on flow rates, SRT is controlled by sludge wasting and recycle, making it a more powerful tool for optimizing bioreactor performance. Since X represents the biomass concentration in the bioreactor, the sludge retention time (SRT) is often denoted by θ_x to emphasize its direct relationship with microbial growth and retention.

In bioreactor design and operation, the HRT (θ) is effectively governed by the SRT (θ_x)—not HRT—controls microbial activity, substrate utilization, and process stability.

Thus:

$$S = \frac{K_s(K_d\theta_x + 1)}{Yq_{\max}\theta_x - K_d\theta_x - 1}$$

$$X = \frac{Y(S^0 - S)}{(K_d\theta_x + 1)}$$

Important trends:

When θ_x is too short, the microbial biomass lacks sufficient time to metabolize the substrate, resulting in effluent substrate concentration (S) approaching the initial substrate concentration (S^0) and biomass concentration (X) approaching zero—a condition known as washout. The critical (minimal) θ_x $\left[\theta_x^{\min}\right]$ occurs precisely at the washout point, where$S = S^0$.

Mathematically, at $\theta_x \rightarrow \theta_x^{\min}$, substrate removal becomes negligible ($S \rightarrow S^0$) and biomass is fully washed out($X \rightarrow 0$). The minimal θ_x is thus defined by$S = S^0$.

By incorporating θ_x into the biomass mass balance equation, the minimal sludge retention time, $\theta_x^{\min}$, can be derived analytically (Rittmann and McCarty 2001).

$$\theta_x^{\min} = \frac{K_s + S^0}{\left(S^0 Y q_{\max} - S^0 K_d - K_d K_s\right)}$$

Further, $\theta_x^{\min}$ increases with increasing initial substrate concentration (S^0) but approaches a theoretical limit when $S \rightarrow \infty$ (or$K_s \rightarrow 0$):

$$[\theta_x^{\min}]_{\text{limit}} = \frac{1}{Y q_{\max} - K_d}$$

As θ_x approaches $\theta_x^{\min}$, the system reaches its lowest sustainable substrate concentration, minimal substrate concentration $\left(S^{\min}\right)$, where residual substrate is just sufficient to maintain a steady-state biomass. This occurs slightly above washout conditions and is given by:

$$S^{\min} = K_s \frac{K_d}{Y q_{\max} - K_d}$$

7.6.4 Mass Balance on Inert Biomass

With the knowledge gained through 'microbial kinetics' and 'steady state mass balance', one can write steady-state mass balance on inert biomass:

$$0 = (1 - f_d) K_d X_a V + Q\left(X_i - X_0^i\right)$$

where

- f_d fraction of the active biomass that is biodegradable
- K_d specific decay constant
- X_a Active biomass in the chemostat $\left(\frac{M_x}{L^3}\right)$
- V volume of chemostat
- X_i concentration of inert biomass in the chemostat $\left(\frac{M_x}{L^3}\right)$
- X_0^i input concentration of inert biomass (or indistinguishable refractory volatile suspended solids) $\left(\frac{M_x}{L^3}\right)$
- Q Flow rate.

From above equation:

$$(1-f_d)K_dX_aV = Q\left(X_i - X_0^i\right)$$

Divide by V

$$\frac{(1-f_d)K_dX_aV}{V} = \frac{Q\left(X_i - X_0^i\right)}{V}$$

$$\text{We know, } \frac{V}{Q} = \theta$$

$$X_i = X_0^i + X_a(1-f_d)K_d\theta$$

The sum of X_i and X_a is called X_v, the volatile suspended solids concentration, or VSS. Letting $\theta_x = \theta$ for the chemostat, X_v can be computed directly as

$$X_v = X_0^i + X_a\left[1 + (1-f_d)K_d\theta_x\right] = X_0^i + Y(S^0 - S)\frac{\left[1 + (1-f_d)K_d\theta_x\right]}{1 + K_d\theta_x}$$

The second term on the right-hand side of above equation represents the net accumulation of biomass from synthesis and decay. It is constituted by the change in substrate concentration ($S^0 - S$) multiplied by the net yield (Y_n).

$$Y_n = Y\frac{\left[1 + (1-f_d)K_d\theta_x\right]}{1 + K_d\theta_x}$$

It is parallel to the relationship between f_s and f_s^0, thus:

$$f_s = f_s^0\frac{\left[1 + (1-f_d)K_d\theta_x\right]}{1 + K_d\theta_x}$$

7.6.5 Activated Sludge Process

The activated sludge process is one of the most important and widely implemented biological wastewater treatment technologies globally. It employs a community of microorganisms—mainly bacteria, but also protozoa and fungi—to degrade organic matter in wastewater. The process occurs in a highly aerated environment that promotes microbial respiration and metabolic activity, thereby converting dissolved and particulate organic pollutants into stable forms such as carbon dioxide, water, and microbial biomass (Grady et al. 2011).

In a typical activated sludge system, wastewater first enters an aeration tank where it is mixed with a concentrated culture of microorganisms, known as the mixed liquor suspended solids (MLSS). The aeration supplies oxygen needed for microbial respiration and keeps the solids in suspension. During this phase, microorganisms metabolize organic compounds in the wastewater, promoting biodegradation. The mixture then flows into a secondary clarifier or settling tank, where the biomass settles by gravity, forming a sludge blanket. A portion of this settled biomass, known as return activated sludge (RAS), is recycled back into the aeration tank to maintain an active microbial population. The remaining sludge, known as waste activated sludge (WAS), is removed from the system for further treatment or disposal (Metcalf and Eddy 2014).

MLSS has further been defined in the next chapter.

The activated sludge process was developed in England in 1914 by Edward Ardern and W. T. Lockett and was first implemented in the United States in the 1920s. Since then, numerous modifications and improvements have been introduced, including extended aeration, contact stabilization, and sequencing batch reactors (SBRs), making the process adaptable to a wide range of wastewater types and flow rates.

This process is highly efficient at removing biochemical oxygen demand (BOD), total suspended solids (TSS), and ammonia nitrogen. It also contributes to phosphorus removal under enhanced biological phosphorus removal (EBPR) configurations. Moreover, the excess sludge produced can be further processed and used as a soil conditioner or fertilizer, closing the loop in resource recovery and environmental sustainability (Tchobanoglous et al. 2003).

7.6.6 *F/M Ratio*

The **Food-to-Microorganism (F/M) ratio** is a critical operational and design parameter in the activated sludge process, used to quantify the balance between the amount of biodegradable organic matter (food) entering the system and the active microbial biomass (microorganisms) responsible for degrading it. It serves as a control metric for reactor stability and treatment efficiency.

The F/M ratio is calculated using the following equation:

$$\frac{\mathrm{F}}{\mathrm{M}} = \frac{\mathrm{Food}\left(\mathrm{BOD\ load}, \frac{\mathrm{lbs}}{\mathrm{day}}\right)}{\mathrm{Microorganisms}(\mathit{MLSS}\ \text{in aeration tank, lbs})}$$

where:

$\mathrm{Food}\left(\frac{\mathrm{lbs}}{\mathrm{day}}\right) = Q(MGD) \times \mathrm{BOD}\left(\frac{\mathrm{mg}}{\mathrm{L}}\right) \times 8.34$

Microorganisms (lbs) $= MLSS \times$ Volume of aeration tank

Food $\left(\frac{\mathrm{lbs}}{\mathrm{day}}\right)$:

The food available to microorganisms is calculated based on the influent flow rate and its biochemical oxygen demand (BOD).

- $\text{Flow(Q)} : \text{in Million Gallons per Day (MGD)}$
- $\text{BOD} : \text{in} \frac{\text{mg}}{\text{L}}$

To convert the product of flow (MGD) and BOD (mg/L) into pounds per day (lbs/day), we use the conversion factor 8.34. This factor represents the weight of one gallon of water in pounds. Therefore:

$$\text{Food}\left(\frac{\text{lbs}}{\text{day}}\right) = \text{Q (MGD)} \times \text{BOD}\left(\frac{\text{mg}}{\text{L}}\right) \times 8.34\left(\frac{\text{lbs}}{\text{day}}\right)$$

Mathematically, the F/M ratio is expressed as:

$$\frac{\text{F}}{\text{M}} = \frac{Q \times S \times 8.34}{V \times X}$$

where:

Q Flow rate (MGD)
S Influent substrate concentration, typically BOD or COD (mg/L)
V Volume of the aeration tank (MG)
X MLSS concentration (mg/L)
8.34 Unit conversion factor (to convert mg/L in MGD to lbs/day)

The F/M ratio is calculated by dividing the weight of the incoming food (measured in pounds of biochemical oxygen demand, or BOD) by the weight of the mixed liquor suspended solids (MLSS) (measured in pounds). The ideal F/M ratio varies depending on the type of wastewater being treated and the desired level of treatment. However, a typical F/M ratio ranges between 0.2 and 0.5 (lb BOD/lb MLSS/day) depending on the process configuration and treatment objectives. Low F/M ratios promote the growth of slow-growing organisms such as nitrifiers and are typical of extended aeration systems, while high F/M ratios are seen in high-rate processes (Rittmann and McCarty 2001).

If the F/M ratio is too high, the microorganisms may experience substrate overload, leading to poor settling characteristics, dispersed growth, and bulking sludge. Conversely, if the F/M ratio is too low, microbial starvation can occur, potentially leading to endogenous respiration and reduced treatment efficiency. Maintaining the F/M ratio within the optimal range is essential for process control and achieving consistent effluent quality.

The F/M ratio is an important parameter to monitor in the activated sludge process. By keeping the F/M ratio within the desired range, operators can ensure that the treatment process is operating efficiently and effectively.

Practice: A wastewater treatment plant receives a daily biochemical oxygen demand (BOD) load of 800 lb. The mixed liquor in the activated sludge system contains

2,000 lb suspended solids (MLSS). Calculate the F/M ratio and determine if the system is operating within the typical range of 0.25 to 0.5.

Solution:

$$\frac{F}{M}\text{ratio} = \frac{800\,\text{lb}}{2000\,\text{lb}} = 0.4$$

The calculated F/M ratio is 0.4, which is within the desired range of 0.25–0.5. This indicates that the balance between food and microorganisms is appropriate for effective treatment.

Practice: A treatment plant is designed to operate at a target F/M ratio of 0.35. If the anticipated BOD load is 700 lb per day, calculate the required amount of MLSS (in pounds) to maintain the desired F/M ratio.

Solution:

$$\frac{F}{M}\text{ratio} = \frac{\text{BOD load (lb)}}{\text{MLSS (lb)}}$$

$$\text{MLSS (lb)} = \frac{\text{BOD load (lb)}}{\frac{F}{M}\text{ratio}} = \frac{700\,\text{lb}}{0.35} = 2000\,\text{lb}$$

To maintain an F/M ratio of 0.35 with an incoming BOD load of 700 lb per day, the system requires 2,000 lb of MLSS.

Further Readings

Doran PM (2013) Bioprocess Engineering Principles", 2nd edn. Academic Press, Waltham, p 2013

Metcalf & Eddy. (2014). *Wastewater Engineering: Treatment and Resource Recovery*, 5th Ed. McGraw-Hill Education.

Egli, T. (1995). The Ecological and Physiological Significance of the Growth of Heterotrophic Microorganisms with Mixtures of Substrates. In: Jones, J.G. (eds) Advances in Microbial Ecology. Advances in Microbial Ecology, vol 14. Springer, Boston, MA.

Grady, C.P.L., Daigger, G.T., Love, N.G., & Filipe, C.D.M. (2011). *Biological Wastewater Treatment, Third Edition.* CRC Press.

Jaibiba P, Naga Vignesh S, Hariharan S (2020) Chapter 10 – Working Principle of Typical Bioreactors. In: Singh L, Yousuf A, Mahapatra DM (eds) Bioreactors. Elsevier, pp 145–173

Rittmann, B.E., McCarty, P.L. (2001). *Environmental Biotechnology: Principles and Applications.* McGraw-Hill.

Tchobanoglous, G., Burton, F.L., Stensel, H.D. (2003). *Wastewater Engineering: Treatment and Reuse.* McGraw-Hill Education.

Chapter 8
Wastewater Treatment—Activated Sludge Process

Learning Objectives:

1. *Understand the stages, components, and operational principles of the activated sludge process in wastewater treatment.*
2. *Apply mass balance and kinetic models to optimize reactor performance using key parameters such as HRT, SRT, MCRT, MLSS, and F/M ratio.*

Wastewater treatment is the process of removing contaminants from wastewater and sewage. The goal of wastewater treatment is to produce a waste stream that is safe to return to the environment, free of toxic pollutants and pathogens. This includes the removal of suspended solids, oils, and organic matter, as well as the treatment of wastewater with chemicals, bacteria, or other processes to neutralize or remove any hazardous compounds.

There are three stages of wastewater treatment:

1. Primary treatment

Primary treatment of wastewater removes settleable, suspended and greasy materials by a physical process. This process includes screens, grit chambers, comminutors, and sedimentation tanks. Screens remove large floating and suspended solids, such as rags and sticks, from the wastewater. Grit chambers consist of a tank that slows down the flow of water and allows the heavier particles, such as sand and gravel, to settle at the bottom. Comminutors grind up large pieces of debris into small pieces. Sedimentation tanks use the force of gravity to separate the solid particles from the wastewater. The solids settle to the bottom and are removed from the wastewater.

2. Secondary treatment

Secondary treatment (also regarded as biological wastewater treatment) involves the removal of biodegradable organic matter (BOD) and total suspended solids (TSS) through the processes of aeration (activated sludge) to activate microorganisms to degrade contaminants and filtration.

V. Achal, *Notes on Biological Processes in Environmental Engineering*,
https://doi.org/10.1007/978-981-95-1610-0_8

3. Tertiary treatment

Tertiary treatment in wastewater treatment is a process of removing pollutants from wastewater to a level that meets the standards for discharge into the environment. In most cases, this involves a physical, chemical, and biological process to remove suspended solids, organic matter, heavy metals, and other pollutants. Tertiary treatment can also involve disinfection such as chlorination, ozonation, or ultraviolet light.

8.1 Activated Sludge Process (ASP)

The activated sludge process is a wastewater treatment process used to remove organic matter from wastewater. It is one of the most widely used biological wastewater treatment processes and is used for treating both domestic and industrial wastewater. The activated sludge process suspended growth process and relies on the use of microorganisms to break down and degrade pollutants in wastewater. The microorganisms feed on the pollutants, using them as a source of energy and breaking them down into simpler, less harmful compounds.

The process begins by mixing the wastewater with a mixture of air and microorganisms known as "activated sludge". This sludge is typically created in a separate tank, known as an "aeration tank", where air is bubbled through the wastewater to provide oxygen to the microorganisms. The microorganisms use the oxygen to break down the organic material in the wastewater, releasing energy and nutrients that they use for growth and reproduction.

The treated wastewater is then sent to a second tank, known as a "settling tank" or "clarifier". Here, the microorganisms and other particles settle to the bottom of the tank. The clarified water is then discharged from the tank, while the sludge is sent to a "sludge digestion" tank. In this tank, the microorganisms break down the organic material further, releasing more energy and nutrients. The sludge is then sent to a "sludge drying bed", where it is dried and can be used as fertilizer or soil conditioner.

The activated sludge process is an effective treatment for removing pollutants from wastewater and is used in many wastewater treatment plants around the world. It is relatively easy to operate and maintain, and it is also relatively cost-effective compared to other treatment processes.

8.2 Three Major Components of ASP (Fig. 8.1)

Aeration tank (also known as activated sludge tank): It is basically bioreactor to grow microorganisms by providing aeration.

Aeration system: It is mechanical system to provide aeration in aeration tank.

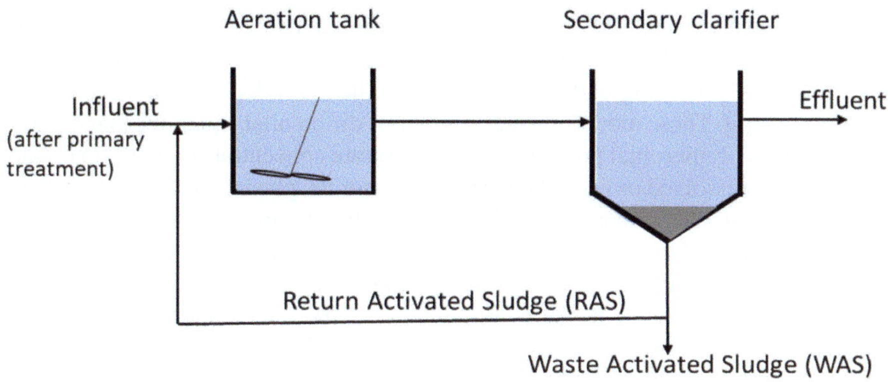

Fig. 8.1 Activated sludge process (ASP)

Secondary clarifier (also known as secondary sedimentation tank or settler): In this tank hungry activated microorganisms/sludge are allowed to settledown, which return to aeration tank or disposed.

8.3 Activated Sludge Process Modelling

Activated sludge process modelling is an essential approach in modern wastewater engineering, enabling the simulation, analysis, and optimization of biological treatment systems (Fig. 8.2). These mathematical models are based on mass balance principles and are used to predict the behavior of activated sludge systems under varying operational and environmental conditions (Henze et al. 2000).

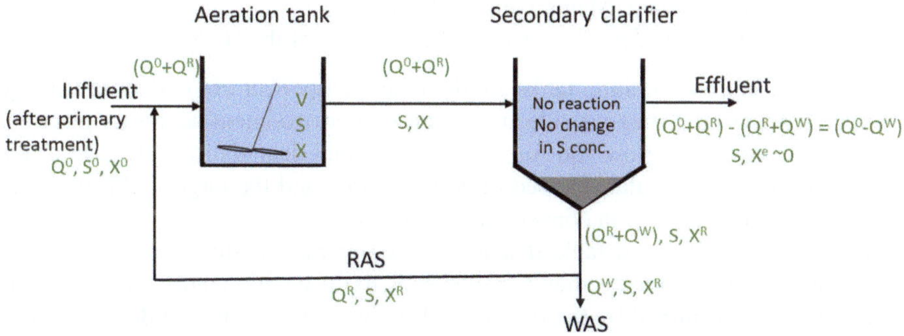

Fig. 8.2 Process modelling of activated sludge (Q^0, S^0, X^0, Q^R, Q^W, X^R, and X^e are initial flow rate, initial substrate concentration, initial biomass concentration, flow rate in RAS, effluent flow rate, biomass concentration in RAS, and biomass concentration in effluent, respectively). RAS = Recycled Activated Sludge, WAS = Waste Activated Sludge

Models such as the ASM (Activated Sludge Models) series—particularly ASM1, ASM2, and ASM3—represent a significant advancement in understanding biological processes, including carbon oxidation, nitrification, denitrification, and phosphorus removal. These models include a series of differential equations that describe the dynamics of microbial populations and substrate concentrations over time. Variables such as dissolved oxygen (DO), ammonia, nitrate, phosphate, chemical oxygen demand (COD), and biomass are modeled with respect to time, aeration, temperature, sludge age, and other factors (Tchobanoglous et al. 2003).

One key application of activated sludge modelling is in the design and scale-up of wastewater treatment plants. By simulating the interactions between operational parameters—such as sludge retention time (SRT), food-to-microorganism (F/M) ratio, and hydraulic retention time (HRT)—engineers and operators can optimize performance, predict effluent quality, and evaluate the impacts of operational changes. In practice, these models are implemented using computational platforms like BioWin, GPS-X, or SIMBA#.

Modelling also plays a vital role in troubleshooting, risk assessment, and compliance monitoring. It provides insights into process stability, microbial population shifts, and treatment resilience under shock loads or toxic inflows. Furthermore, models assist in the transition toward resource recovery by predicting nutrient removal and energy generation potential (Daigger 2011).

8.4 Hydraulic Retention System (HRS)

A Hydraulic Retention System (HRS) is designed to regulate the hydraulic retention time (HRT), which is defined as the average time a unit volume of wastewater remains in the treatment system. In the context of the activated sludge process, the HRT determines the contact time between wastewater and microorganisms, influencing biodegradation efficiency (Metcalf and Eddy 2014).

Several components contribute to the effectiveness of the HRS:

- **Aeration Tanks**: Aeration tanks are the main component of an activated sludge treatment system. They provide an oxygen-rich environment necessary place for microorganisms to grow and for aerobic microbial degradation. The size of the aeration tank is calculated based on the flow rate and the target HRT, typically ranging from 6 to 12 h in conventional systems.
- **Clarifiers**: Clarifiers are tanks that are used to separate the sludge from the treated wastewater. The sludge is then recycled back to the aeration tank. The size of the clarifier is determined by the flow rate of the wastewater and the desired sludge loading rate.
- **Sludge Thickeners**: These increase the solid content of waste activated sludge before dewatering. Effective thickening optimizes subsequent sludge handling and disposal.

Maintaining proper HRT through a well-designed HRS ensures effective treatment, process stability, and compliance with effluent standards. Improper HRT can lead to incomplete treatment, washout of biomass, or excessive sludge production.

8.5 Sludge Age and Solid Retention Time (SRT, θ_x)

Sludge age and solid retention time (SRT, denoted θ_x) are fundamental operational parameters that define the average time microorganisms or solids remain in the activated sludge system. While sludge age focuses on microbial population dynamics, SRT includes all suspended solids, including inert and non-biodegradable fractions.

The SRT influences the biological treatment capacity, particularly in nutrient removal processes. A longer SRT favors nitrifying bacteria, which have slower growth rates, and enhances the degradation of complex organics. Conversely, short SRTs may improve sludge settleability but limit nitrification and lead to higher sludge production.

The optimal SRT depends on the specific treatment goal. Typical values range from 5–10 days for conventional BOD removal, and 15–25 days for nutrient removal systems.

Sludge age and SRT can be controlled by adjusting the rate of sludge wasting. Sludge wasting is the process of removing excess sludge from the activated sludge system. The rate of sludge wasting is typically controlled by monitoring the concentration of mixed liquor suspended solids (MLSS).

By controlling sludge age and SRT, operators can ensure that the activated sludge process is operating efficiently and effectively.

$$\theta_x = \frac{\text{Active biomass in the system}}{\text{Production rate of active biomass}}$$

From Fig. 8.2:

$$\theta_x = \frac{XV}{Q_{\text{effluent}}X_{\text{effluent}} + Q_{WAS}X_{WAS}}$$

8.6 Mixed Liquor Suspended Solids (MLSS)

Mixed liquor suspended solids (MLSS) is the concentration of suspended solids in mixed liquor, usually expressed in milligrams per liter (mg/L). Mixed liquor is a mixture of raw or settled wastewater and activated sludge contained in an aeration tank in the activated sludge process. MLSS consists mostly of microorganisms and non-biodegradable suspended matter.

MLSS is important because it is a measure of the biomass of microorganisms in the activated sludge system. MLSS is managed by adjusting sludge wasting and aeration control. High MLSS may lead to oxygen transfer limitations and poor settleability, while low MLSS may reduce treatment efficiency. Regular monitoring is essential for process control (Tchobanoglous et al. 2003).

Here are some of the benefits of maintaining a good MLSS concentration:

- It helps to ensure that there are enough microorganisms present to remove pollutants from the wastewater.
- It helps to prevent sludge bulking, which is a condition in which the sludge becomes too thick and settles poorly.
- It helps to control the release of nutrients into the wastewater, which can promote the growth of algae and other undesirable organisms.

Here are some of the risks of having too high or too low of an MLSS concentration:

- A too high MLSS concentration can lead to sludge bulking, which can reduce the efficiency of the activated sludge process.
- A too low MLSS concentration can lead to a decrease in the number of microorganisms present, which can also reduce the efficiency of the activated sludge process.

Typical MLSS concentrations range from 2,000 to 4,000 mg/L in conventional systems, though they may be higher in advanced treatment processes such as membrane bioreactors (MBRs).

8.7 Mean Cell Residence Time (MCRT)

Mean cell residence time (MCRT) is the average time that a microorganism spends in the activated sludge process. It is calculated by dividing the total mass of microorganisms in the system by the mass of microorganisms that are wasted each day. MCRT is typically expressed in days.

MCRT is an important parameter in the activated sludge process because it affects the performance of the process. A longer MCRT will help to remove more pollutants from the wastewater, but it will also produce more sludge. A shorter MCRT will produce less sludge, but it may not remove as many pollutants from the wastewater.

$$\text{MCRT(days)} = \frac{\text{Total suspended solids (TSS) in the system}}{\text{TSS removed daily}}$$

The MLSS consists of all solids in the aeration tanks and secondary clarifiers, represents TSS in the system. Total suspended solids leaving the activated sludge process consists of the amount of suspended solids loss through wasting and discharge in the secondary effluent. Thus, MCRT can be formulated as:

$$\text{MCRT} = \frac{\text{MLSS (Volume of Aeration tank + Volume of Clarifier)}}{\text{MLSS leaving secondary system}(WAS + \text{Effluent})}$$

$$MCRT = \frac{(XV_{\text{Aeration tank}} + XV_{\text{Clarifier}})}{(Q^e X^e + Q^w X^w)}$$

where:

X $isMLSS\left(\frac{mg}{L}\right)$
Q^e, X^e effluent flow and solids
Q^w, X^w waste sludge flow and solids

There are other equations that can be derived from Fig. 8.2, important in solving mathematical problems, summarized in Table 8.1.

MCRT plays a vital role in biological nutrient removal, particularly for systems targeting nitrification and phosphorus uptake. A longer MCRT (>10 days) is generally required for stable nitrification. Short MCRT values may cause washout of slow-growing autotrophs, leading to elevated ammonia levels in the effluent (Rittmann and McCarty 2001).

Practice:

At an activated sludge wastewater treatment receiving 3.25 MGD (Millions of Gallons per Day), the final effluent suspended solids concentration averages 21.2 mg/L. what would the calculated MCRT value be when the aeration tank carries 2050 mg/L MLSS and wastes 0.055 MGD. The waste activated sludge (WAS) has a concentration of 7980 mg/L. The aeration tank has a volume of 1 MG (Million Gallons) and the secondary clarifier has an operational volume of 0.25 MG (Million Gallons).

Table 8.1 Chart of major equations useful for a CSTR with biomass settling and recycling at steady state

Parameter	Formula
Hydraulic Retention Time (HRT)	$\theta = \frac{V}{Q}$
Mean Cell Residence Time (MCRT)	$\text{MCRT} = \frac{XV_{\text{Aeration tank}} + XV_{\text{Clarifier}}}{Q^e X^e + Q^w X^w}$
Solid Retention Time (SRT)	$\theta_x = \frac{XV}{Q^e X^e + Q^w X^w}$
SRT at which biomass wash out	$\theta_x^{\min} = \frac{K_s + S^0}{(S^0 Yq_{\max} - S^0 K_d - K_d K_s)} S \to S^0$
SRT limit at which biomass wash out	$\left[\theta_x^{\min}\right]_{\lim} = \frac{1}{Yq\max - K_d} S \to \infty$
Effluent substrate concentration	$S = K_s \frac{(K_d \theta_x + 1)}{(Yq\max - K_d \theta_x - 1)}$
Minimal effluent substrate concentration	$S^{\min} = K_s \frac{Kd}{Yq\max - K_d} \theta_x \to \infty$
Biomass concentration	$X = \frac{\theta_x}{\theta} \frac{Y(S^0 - S)}{1 + K_d \theta_x}$
Biomass net production rate	$r_x = \frac{XV}{\theta_x}$
Net yield	$Y_{\text{net}} = \frac{Y}{1 + K_d \theta_x}$

Solution:

Given values–

Plant flow (Q^e) = 3.25 MGD.

Final effluent SS (X^e) = 21.2 mg/L.

WAS Flow (Q^w) = 0.055 MGD.

WAS SS (X^w) = 7980 mg/L.

Aeration tank volume ($V_{\text{Aeration tank}}$) = 1 MG.

Clarifier volume ($V_{\text{Clarifier}}$) = 0.25 MG.

MLSS (X) = 2050 mg/L.

We know: 8.34 is the weight of one gallon of water in pounds at room temperature.

$$\text{MCRT} = \frac{(XV_{\text{Aeration tank}} + XV_{\text{Clarifier}})}{(Q^e X^e + Q^w X^w)}$$

$$\text{MCRT} = \frac{(1\,\text{MG} + 0.25\,\text{MG}) \times \left(2050\frac{\text{mg}}{\text{L}}\right) \times (8.34)}{\left[(3.25\,\text{MGD})\left(21.2\frac{\text{mg}}{\text{L}}\right)(8.34) + (0.055\,\text{MGD})\left(7980\frac{\text{mg}}{\text{L}}\right)(8.34)\right]} = 5\,\text{days}$$

8.7.1 Engineering Design of Reactors

When designing bioreactors for water and wastewater treatment, engineers must ensure not only that the systems achieve the required levels of contaminant removal but also that they operate reliably under a variety of conditions. Much of the design work relies on mathematical models that relate substrate removal and biomass production to key operational parameters such as hydraulic detention time and solids retention time. However, the performance specifications—whether expressed in terms of percentage removal (for example, an 85% reduction in BOD) or absolute effluent quality limits (such as effluent concentrations not exceeding 30 mg/L for BOD and suspended solids)—form only one part of the design challenge.

Simply designing a system to meet these performance levels under ideal conditions is not sufficient. Just as structural engineers use a factor of safety to ensure that buildings and bridges can support more than the expected loads, environmental engineers must incorporate safety factors into reactor design. This extra margin accounts for inevitable variations in influent composition, unexpected operational disruptions, and the natural variability of biological processes.

A practical method for applying a safety factor involves setting the design parameter—as an example, the effective reaction rate or substrate loading—to a value that is a multiple of the most critical or limiting measured value. In mathematical terms, one might express this as:

$$\text{Design Parameter} = \text{Measured Limiting Value} \times \text{Safety Factor(SF)}$$

Here, the safety factor (SF) is chosen based on empirical data collected over years of operational experience. For instance, early designs of suspended-growth reactors, like those used in aerobic activated sludge systems, evolved largely through trial and error. Over time, standardized safety factors have emerged from this experience, ensuring that reactor designs are robust and reliable even when actual conditions deviate from ideal laboratory conditions.

One recommended method for incorporating a safety factor in treatment reactor design—based on the work of Christensen and McCarty (1975)—is to determine the design parameter (θ_x) as a multiple of its minimum limiting value. In this approach, the design parameter is calculated by multiplying the minimal value by a safety factor (SF), as expressed by the formula:

$$\theta_x^d = SF \times \left[\theta_x^{\min}\right] \lim$$

In simple way:

$$SF = \frac{\theta x}{\theta_x^{\min}}$$

In short, by deliberately incorporating a safety factor, engineers design treatment plants that exceed the minimum regulatory performance requirements. This approach guarantees an operational buffer, ensuring that the reactor continues to perform effectively under variable conditions and throughout its service life. Balancing design efficiency with operational reliability is critical for environmental systems, where the primary goal is to safeguard public health and environmental quality over the long term. All the important formula being used to calculate various mathematical problems in CSTR are summarized in Table 8.1.

Practice: A 2,000 m^3 chemostat receives 1,000 m^3/d wastewater with soluble biodegradable organic pollutant concentration of 500 mg BOD/L. Kinetic parameters are given below:

$$Ks = 20\,\text{mg}\frac{\text{BOD}}{\text{L}},\ Kd = 0.15\frac{1}{d},\ q_{\max} = 20\frac{\text{gBOD}}{\text{gVSS}} \cdot d$$

$Y = 0.42\frac{\text{g VSS}}{\text{g BOD}}$. Assume the reaction is aerobic.
Find out:

(a) What is the sludge age and minimum value of the mean cell residence time?
(b) What is the pollutant concentration at the effluents?
(c) What is pollutant removal efficiency?
(d) What is the biomass concentration?

Solution:

(a) For such chemostat, SRT (θ_x) would be equal to HRT (θ). Thus,

$$\theta_x = \theta = \frac{V}{Q} = \frac{2000\,\mathrm{m}^3}{1000\,\mathrm{m}^3/\mathrm{d}} = 2\,\mathrm{days}$$

Sludge age = 2 days.
Minimal value of mean cell residence time ($\theta_x^{\min}$)

$$\theta_x^{\min} = \frac{K_s + S^0}{\left(S^0 Y q_{\max} - S^0 K_d - K_d K_s\right)}$$

$$= \frac{20\,\mathrm{mg}\frac{\mathrm{BOD}}{\mathrm{l}} + 500\,\mathrm{mg}\frac{\mathrm{BOD}}{\mathrm{L}}}{\left(500\,\mathrm{mg}\frac{\mathrm{BOD}}{\mathrm{L}} \times 0.42\frac{\mathrm{gVSS}}{\mathrm{gBOD}} \times 20\frac{\mathrm{gBOD}}{\mathrm{gVSS{\cdot}d}} - 50\,\mathrm{mg}\frac{\mathrm{BOD}}{\mathrm{L}} \times 0.15\frac{1}{d} - 20\,\mathrm{mg}\frac{\mathrm{BOD}}{\mathrm{L}} \times 0.15\frac{1}{\mathrm{d}}\right)}$$

$$= 0.126\,\mathrm{days}$$

(b) Pollutant concentration at the effluents (S)

$$S = Ks\frac{(Kd\theta_x + 1)}{(Yq_{\max}\theta_x - Kd\theta_x - 1)}$$

$$= 20\,\mathrm{mg}\frac{\mathrm{BOD}}{\mathrm{L}}\frac{\left(0.15\frac{1}{\mathrm{d}} \times 2\mathrm{d} + 1\right)}{\left(0.42\frac{\mathrm{gVSS}}{\mathrm{gBOD}} \times 20\frac{\mathrm{gBOD}}{\mathrm{gVSS{\cdot}d}} \times 2\mathrm{d} - 0.15\frac{1}{\mathrm{d}} \times 2\mathrm{d} - 1\right)}$$

$$= 1.7\,\mathrm{mg}\frac{\mathrm{BOD}}{\mathrm{L}}$$

(c) Pollutant removal efficiency (RE)

$$RE = \frac{\left(S^0 - S\right)}{S^0} \times 100\%$$

$$RE = \frac{\left(500\,\mathrm{mg}\frac{\mathrm{BOD}}{\mathrm{L}} - 1.7\,\mathrm{mg}\frac{\mathrm{BOD}}{\mathrm{L}}\right)}{500\,\mathrm{mg}\frac{\mathrm{BOD}}{\mathrm{L}}} \times 100\% = 99.66\%$$

(d) Biomass concentration (X)

$$X = \frac{Y\left(S^0 - S\right)}{(K_d\theta_x + 1)}$$

$$= \frac{0.42\frac{\mathrm{gVSS}}{\mathrm{gBOD}}(500 - 1.7)\,\mathrm{mg}\frac{\mathrm{BOD}}{\mathrm{l}}}{\left(0.15\frac{1}{\mathrm{d}} \times 2\mathrm{d} + 1\right)} = 161\frac{\mathrm{mgVSS}}{\mathrm{L}}$$

Practice:

At an activated sludge wastewater treatment plant receiving 4.5 L/day, the final effluent suspended solids concentration averages 18.5 mg/L. What would the calculated Solids Retention Time (SRT) value be when the aeration basin contains 2500 mg/L Mixed Liquor Suspended Solids (MLSS) and wastes 0.0650 L/day? The Waste Activated Sludge (WAS) has a concentration of 8200 mg/L. The aeration basin has a volume of 1.2 L, and the secondary clarifier has an operational volume of 0.300 L.

Solution:

To calculate the SRT, θ_x, we use the formula:

$$\theta_x = \frac{XV}{Q^e X^e + Q^w X^w}$$

where:

- V volume of the aeration basin
- X concentration of MLSS in the aeration basin
- Q^w waste activated sludge flow rate
- X^w concentration of solids in the waste activated sludge
- Q^e effluent flow rate (assuming it's equal to the influent flow rate minus the waste flow rate, since there's usually no significant additional water removal)
- X^e concentration of suspended solids in the effluent

Given values:

- V 1.2 L (aeration basin volume)
- X 2500 mg/L (MLSS concentration)
- Q^w 0.0650 L/day (WAS flow rate)
- X^w 8200 mg/L (WAS concentration)
- Q^e $4.5 - 0.0650 = 4.435$ L/day (effluent flow rate)
- X^e 18.5 mg/L (effluent suspended solids concentration)

Let's calculate the SRT value based on the given data.

$$\theta_x = \frac{2500 \times 1.2}{4.435 \times 18.5 + 0.065 \times 8200} = 4.88 \text{ days}$$

The calculated Solids Retention Time (SRT) value for the given scenario is approximately 4.88 days. This indicates the average time that the solids are retained in the system, which is crucial for ensuring effective treatment and biological breakdown of waste in the activated sludge process.

Further Readings

Christensen DR, McCarty PL (1975) Multi-process biological treatment model. Journal of the Water Pollution Control Federation 47:2652–2664

Daigger GT (2011) A practitioner's perspective on the application of wastewater treatment modelling. Water Sci Technol 63(3):516–526

Metcalf & Eddy. (2014). *Wastewater Engineering: Treatment and Resource Recovery*, 5th Ed. McGraw-Hill.

Henze, M., Gujer, W., Mino, T., & van Loosdrecht, M. C. M. (2000). *Activated sludge models ASM1, ASM2, ASM2d and ASM3.* (reprint ed.) IWA Publishing.

Rittmann, B.E., & McCarty, P.L. (2001). *Environmental Biotechnology: Principles and Applications.* McGraw-Hill.

Tchobanoglous, G., Burton, F.L., & Stensel, H.D. (2003). *Wastewater Engineering: Treatment and Reuse.* McGraw-Hill Education.

Chapter 9
Advanced Wastewater Treatment

Learning Objectives:

1. *Understand the advanced biological processes and technologies used for nutrient removal—including nitrification, denitrification, and biological phosphorus removal—in wastewater treatment.*
2. *Apply mass balance and kinetic modeling principles to design, optimize, and evaluate advanced treatment systems such as the Modified Ludzack-Ettinger process, post-anoxic denitrification, and the PhoStrip process.*

Advanced wastewater treatment (AWT) is any process that removes impurities from wastewater beyond what is attainable through conventional secondary or biological treatment. AWT is used to remove nutrients, such as nitrogen and phosphorus, and other pollutants, such as heavy metals, pharmaceuticals, and endocrine disruptors. AWT is often used to treat wastewater from industries and municipalities that have high levels of pollutants.

Nitrogen removal is a critical part of wastewater treatment. Nitrogen is an essential nutrient for plants and animals, but too much nitrogen can be harmful to the environment. When nitrogen is released into waterways, it can cause eutrophication, which is a condition in which excessive amounts of nutrients cause algae to grow rapidly. This can lead to the formation of harmful algal blooms, which can deplete oxygen levels in the water, killing fish and other aquatic life.

There are a number of different methods for removing nitrogen from wastewater. One common method is nitrification, which is a biological process that converts ammonia (NH_3) into nitrite (NO_2^-) and then nitrate (NO_3^-). Nitrifying bacteria use ammonia as a source of nitrogen and oxygen as an electron acceptor. The nitrification process is carried out in two steps:

Ammonia oxidation: Ammonia is oxidized to nitrite by ammonia oxidizing bacteria (AOB) such as *Nitrosomonas*, *Nitrosococcus*, and *Nitrobacter*.

Nitrite oxidation: Nitrite is oxidized to nitrate by nitrite oxidizing bacteria (NOB) such as *Nitrospira* and *Nitrococcus*.

V. Achal, *Notes on Biological Processes in Environmental Engineering*,
https://doi.org/10.1007/978-981-95-1610-0_9

Once the nitrogen has been converted to nitrate, it can be removed from wastewater using a number of different methods, including denitrification.

Denitrification is another biological process that converts nitrate into nitrogen gas (N_2). This process is carried out by denitrifying bacteria, which use nitrate as a terminal electron acceptor in the absence of oxygen. Denitrification is typically carried out in anoxic conditions, which means that there is no free oxygen present. Anoxic conditions can be created by adding an inert gas, such as carbon dioxide, to the wastewater.

9.1 Nitrification

Nitrification is a critical, two-step microbial process that transforms ammonia (NH_3) into nitrate (NO_3^-) via an intermediate nitrite (NO_2^-). In the first phase, ammonia-oxidizing bacteria (AOB) oxidize ammonia to form nitrite. In the second phase, nitrite-oxidizing bacteria (NOB) convert nitrite into nitrate.

Both AOB and NOB are autotrophic, meaning they derive their energy from inorganic sources. They oxidize ammonia and nitrite, respectively, using oxygen as the terminal electron acceptor. The energy released during these reactions supports bacterial growth and cell maintenance.

Because nitrification requires oxygen, it is strictly an aerobic process. Inadequate dissolved oxygen can hinder or even stop nitrification, a concern that often arises when wastewater is too concentrated or if the treatment plant is not operated correctly.

Nitrification plays a vital role in the nitrogen cycle—a series of processes that convert nitrogen among its various chemical forms in the environment. Despite the abundance of atmospheric nitrogen, plants and animals rely on bioavailable forms such as nitrate. Once ammonia is oxidized to nitrate, plants absorb it through their roots to synthesize proteins and other essential compounds.

Beyond nutrient cycling, effective nitrification also helps prevent water pollution. Excessive ammonia in water bodies can trigger eutrophication, a process that stimulates excessive algal growth. When these algal blooms decompose, they can severely deplete dissolved oxygen, harming fish and other aquatic organisms.

Several factors can influence the nitrification process:

- **Temperature**: As an enzyme-driven process, nitrification is most efficient between 20 and 30 °C.
- **pH**: Nitrifying bacteria favor slightly alkaline conditions, with an optimal pH range of 7.5–8.5.
- **Oxygen**: Adequate oxygen is essential; insufficient levels can slow or inhibit the oxidation reactions.
- **Nutrients**: The availability of nutrients such as carbon, nitrogen, and phosphorus is necessary to support the growth and activity of nitrifying bacteria.

Understanding these principles is essential for designing and operating wastewater treatment systems that effectively support nitrification, thereby maintaining water quality and protecting the environment.

Practice: A treatment plant was designed to carry out nitrification, as well as BOD removal, in a one-sludge system. In order to assure that nitrification occurred, the designers used a low loading rate of 0.2 kg BODs/m^3 d. Nonetheless, nitrification perfom1ance has been poor. You are the high-priced consultant and have obtained the following operating data: water temperature: 15 °C, aeration basin D.O. = 2.5 mg/L, aeration basin regime: essentially completely mixed, MLVSS = 1,200 mg/L, Q = 10^4 m^3/d, reactor volume = 1.25×10^4 m^3.

- Influent quality: BODs = 250 mg/L, total Kjeldahl N = 23 mg/L
- Effluent qUality: BODs = 25 mg/L, VSS = 25 mg/L, $NH_4^+ - N = 20$ mg N/L
- Recycle sludge quality: recycle ratio = 0.32, volatile solids content = 5×10^3 mg/L
- Waste sludge flow rate: 9×10^2 m^3/d

What is the problem and how can it be solved?

Solution:

Key observations from data:

- **High Ammonia in Effluent**: With influent nitrogen 23 mg/L and effluent ammonia 20 mg/L, very little ammonia is converted.
- **Operating Conditions**: At 15 °C, nitrifying bacteria grow slowly compared to heterotrophs.
- **Sludge Management Data**
  - The aeration basin has MLVSS = 1,200 mg/L.
  - The reactor volume is large (1.25×10^4 m^3) while the flow rate is 10^4 m^3/d.
  - The wasted sludge (waste flow of 9×10^2 m^3/d) is taken from a stream with a volatile solids concentration of 5×10^3 mg/L.
  - A recycle ratio of 0.32 indicates that a significant fraction of the biomass is removed as waste rather than recycled to the aeration basin.
- **Reactor volume**: 1.25×10^4 m^3
- **Mixed Liquor Volatile Suspended Solids (MLVSS)**: 1,200 mg/L

Convert reactor volume to liters:

$$1.25 \times 10^4\,\mathrm{m^3} \times 1000\,\frac{\mathrm{L}}{\mathrm{m^3}} = 1.25 \times 10^7\,\mathrm{L}$$

Mass of biomass (in mg):

$$1.25 \times 10^7\,\mathrm{L} \times 1200\frac{\mathrm{mg}}{\mathrm{L}} = 1.5 \times 10^{10}\,\mathrm{mg}$$

Convert to kilograms:

$$\frac{1.5 \times 10^{10}\,\text{mg}}{1 \times 10^{6}} = 15000\,\text{Kg}$$

- **Waste sludge flow rate:** 900 m^3/d

Convert waste flow to liters:

$$900\,\frac{\text{m}^3}{\text{d}} \times 1000 = 900,000\frac{\text{L}}{\text{d}}$$

- **Wasted sludge concentration:** 5,000 mg/L

Mass of waste biomass (in mg/d):

$$900,000\frac{\text{L}}{\text{d}} \times 5000\frac{\text{mg}}{\text{L}} = 4.5 \times 10^{9}\frac{\text{mg}}{\text{d}}$$

Convert to kilograms per day:

$$\frac{4.5 \times 10^{9}\,\frac{\text{mg}}{\text{d}}}{1 \times 10^{6}} = 4500\frac{\text{Kg}}{\text{d}}$$

SRT Calculation

$$SRT = \frac{XV}{Q^{w}X^{w}}$$

where:

X $MLSS\left(\frac{mg}{L}\right)$
V Reactor volume (L)
Q^{w} Waste flow rate $\left(\frac{L}{d}\right)$
X^{w} Concentration in the waste sludge $\left(\frac{\text{mg}}{\text{L}}\right)$

Using the provided data:

V 1.25×10^{7} L
X 1,200 mg/L
Q^{w} 900,000 L/d
X^{w} 5,000 mg/L

$$\text{SRT} = \frac{1.25 \times 10^{7} \times 1200}{900,000 \times 5000} = 3.33\,\text{days}$$

Impact on Nitrification:

Nitrifiers require a minimum SRT (SRT_{min}) that is higher than the SRT of heterotrophs, especially under low-temperature conditions (15 °C). Typical SRT_{min} for nitrifiers at 15 °C is in the range of 8–10 days or more. When:

$$\text{SRT(actual)} < SRT_{min} \Rightarrow \text{Nitrifier washout, poor ammonia oxidation.}$$

The plant's nitrification performance is poor because the operating SRT is far too short (approximately 3.33 days) to support the slow-growing nitrifying bacteria at 15 °C. As a consequence, nearly all influent ammonia (NH_4^+–N) remains in the effluent (20 mg/L vs. 23 mg/L influent total Kjeldahl N).

To improve nitrification, the SRT must be increased. This can be achieved by reducing the waste sludge flow rate (or reducing the solids concentration in the waste sludge) so that the reactor retains biomass long enough for nitrifiers to proliferate (targeting an SRT of at least 8–10 days). Additionally, process modifications (e.g., implementing selective wasting or a two-zone reactor design) can help further ensure that nitrifying bacteria are not washed out. By making these adjustments, the slow-growing nitrifiers will have sufficient time to grow, leading to improved ammonia conversion and overall nitrification performance.

9.2 Denitrification

Denitrification is a microbially driven process in which nitrate (NO_3^-) is sequentially reduced to nitrogen gas (N_2) through a series of intermediate nitrogen oxide compounds. This pathway is predominantly carried out by facultative anaerobic bacteria that shift to using nitrate as an electron acceptor when oxygen is scarce, utilizing organic matter or other suitable electron donors to drive the reactions. As such, denitrification serves as a complementary process to aerobic nitrification within the broader nitrogen cycle, playing a vital role in removing excess nitrogen from various environments. Denitrification is carried out by a variety of bacteria, including *Pseudomonas*, *Paracoccus*, and *Thiobacillus*. These bacteria use nitrate as an electron acceptor, in place of oxygen, during respiration. This process releases energy, which the bacteria use to grow and reproduce.

In this anaerobic process, bacteria first convert nitrate to nitrite via the enzyme nitrate reductase. This is followed by a stepwise reduction where nitrite is transformed into nitric oxide (NO), then nitrous oxide (N_2O), and finally into molecular nitrogen (N_2). Each of these transformations is facilitated by specific enzymes tailored to operate under oxygen-deficient conditions. The intermediate products, particularly nitrous oxide, are released as gases during the conversion process.

Denitrification is critical not only for maintaining environmental nitrogen balance but also for preventing issues associated with nutrient over-enrichment, such as eutrophication. Excess nitrogen in aquatic ecosystems can lead to rapid algal proliferation, subsequent oxygen depletion, and ultimately, adverse impacts on aquatic life.

By converting nitrate to nitrogen gas, denitrification helps mitigate these negative effects.

This process is not limited to natural settings; it is also harnessed in engineered systems like wastewater treatment plants. In such facilities, denitrification is strategically implemented to remove nitrates from effluents before discharge, thereby reducing the risk of downstream environmental damage.

Several key factors influence the efficiency of denitrification:

- **Oxygen Concentration**: Since denitrification is strictly an anaerobic process, the presence of oxygen can inhibit the activity of denitrifying bacteria.
- **Nitrate Availability**: Adequate nitrate must be present for denitrification to proceed, as it serves as the primary electron acceptor.
- **Organic Carbon**: The availability of organic matter is essential, providing the electrons needed for the reduction of nitrate.
- **Temperature**: The enzymatic processes involved typically function best within a moderate temperature range, generally between 20 and 30 °C.
- **pH**: Denitrification is most efficient under slightly acidic to neutral conditions, with an optimal pH range of 6.5–7.5.

Understanding and managing these factors are crucial for the successful application of denitrification in wastewater treatment. By optimizing conditions to support denitrifying bacteria, operators can ensure that the process consistently reduces nitrate levels and maintains overall treatment efficiency.

9.3 Biological Nitrogen Removal Process

There are three processes commonly used for removing nitrogen from wastewater: modified Ludzack-Ettinger process (MLE), post anoxic denitrification process, and two-sludge system process.

9.3.1 Modified Ludzack-Ettinger Process (MLE)

The modified Ludzack-Ettinger process (MLE) is a biological process for removing nitrogen from wastewater. It is a variation of the activated sludge process, and it uses a combination of aerobic and anoxic conditions to promote the growth of nitrifying and denitrifying bacteria.

The Modified Ludzack-Ettinger (MLE) process is a widely used biological nitrogen removal configuration that enhances the conventional activated sludge process by integrating both aerobic and anoxic zones. The MLE configuration promotes effective nitrification and denitrification within a single-sludge system and is especially suited for municipal and industrial wastewaters with moderate to high organic content.

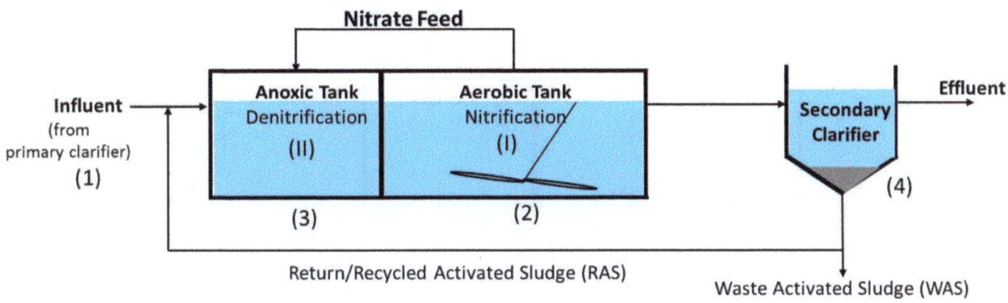

Fig. 9.1 Schematic representation of Modified Ludzack-Ettinger (MLE) process

As illustrated in Fig. 9.1, in the MLE process the flow begins in the aerobic zone, where nitrifying bacteria oxidize ammonia (NH_4^+) first to nitrite (NO_2^-) and then to nitrate (NO_3^-) under the influence of dissolved oxygen. A portion of this nitrate-rich mixed liquor is then internally recycled to a preceding anoxic zone, where denitrifying bacteria reduce nitrate to nitrogen gas (N_2) using the readily biodegradable organic carbon available in the influent.

This configuration allows the nitrate generated during nitrification to serve as an electron acceptor in the upstream anoxic zone, where organic matter is still abundant—eliminating or minimizing the need for external carbon dosing.

The process is designed to exploit the complementary metabolic needs of two distinct microbial populations:

- Nitrifying bacteria, which require oxygen to oxidize ammonia (NH_4^+) to nitrate (NO_3^-)
- Denitrifying bacteria, which function under anoxic conditions to reduce nitrate to nitrogen gas (N_2)

The MLE process is a more efficient way to remove nitrogen from wastewater than the conventional activated sludge process. This is because the MLE process takes advantage of the different metabolic requirements of nitrifying and denitrifying bacteria. It is also more energy-efficient than the conventional activated sludge process. This is because the aeration requirements are lower in the anoxic stage. Further, the MLE process is a versatile process that can be used to treat a variety of wastewaters. It is particularly well-suited for treating wastewater with high concentrations of organic matter.

9.3.2 Post Anoxic Denitrification

Post anoxic denitrification is a biological process that uses denitrifying bacteria to reduce nitrate (NO_3^-) to nitrogen gas (N_2) in wastewater treatment systems. It is a variation of the traditional denitrification process, which requires an anaerobic environment.

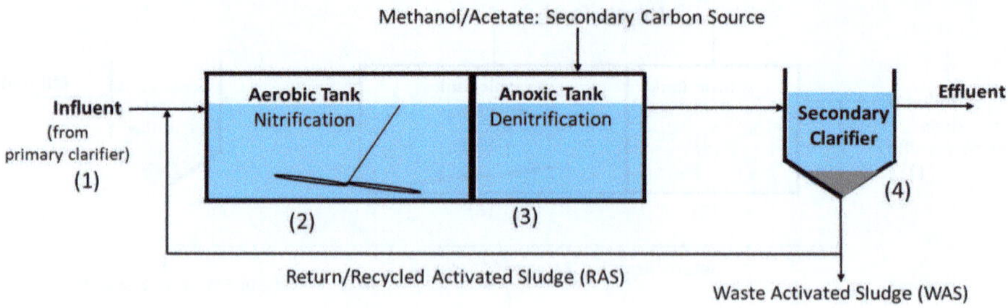

Fig. 9.2 Schematic representation of the Post Anoxic Denitrification process

In post anoxic denitrification, the treatment process begins in an aerobic zone, where nitrifying bacteria convert ammonia (NH_4^+) to nitrite (NO_2^-), and subsequently to nitrate (NO_3^-). The now nitrified wastewater is directed to an anoxic zone, where denitrifying bacteria utilize nitrate as an electron acceptor in the absence of oxygen, reducing it to nitrogen gas (N_2), which is released harmlessly into the atmosphere (Fig. 9.2). The process generally requires the presence of a biodegradable carbon source, which is either provided by residual organic matter in the influent or through external carbon addition (e.g., methanol, acetate) (Cadmus Group, Inc. 2010).

Unlike pre-anoxic systems where the carbon source in the influent is used for denitrification before nitrification, post anoxic systems often face carbon limitation, since most of the organic matter is already degraded during the aerobic step. Therefore, external carbon dosing is a key design consideration to ensure complete nitrate reduction.

The post anoxic denitrification process has several advantages over the traditional denitrification process. First, it is more efficient because it takes advantage of the different metabolic requirements of nitrifying and denitrifying bacteria. Second, it is more energy-efficient because the aeration requirements are lower in the anoxic zone. Third, it is more versatile because it can be used to treat a wider range of wastewater.

9.3.3 Two-Sludge System Process

The two-sludge system process is an advanced biological treatment method designed specifically for the removal of nitrogen from wastewater. As an evolution of the conventional activated sludge process, this approach employs two distinct sludge populations to optimize the conditions for nitrifying and denitrifying bacteria, thereby enhancing overall nitrogen removal efficiency.

The two-sludge process consists of three stages:

Aerobic Stage: In this stage, the wastewater is vigorously aerated, creating an environment that supports the growth of nitrifying bacteria. These microorganisms oxidize ammonia (NH_3) first into nitrite (NO_2^-) and then into nitrate (NO_3^-). The production of nitrate during this phase is critical, as it serves as the substrate for subsequent denitrification.

Denitrification Stage: Following the aerobic process, the wastewater is conveyed to an anoxic zone where denitrifying bacteria take center stage. In the absence of oxygen, these bacteria reduce nitrate to nitrogen gas (N_2), using organic carbon as an electron donor. This conversion effectively removes nitrate from the water, minimizing the risk of nitrogen-related environmental impacts.

Sludge Separation Stage: The final stage involves the separation of biomass from the treated wastewater. Here, the system distinguishes between the two types of sludge: the nitrifying sludge, which is recycled back to the aerobic stage to continue processing ammonia, and the denitrifying sludge, which is returned to the denitrification stage to sustain nitrate reduction.

By utilizing separate sludge streams, the two-sludge system process effectively harnesses the specific metabolic capacities of nitrifying and denitrifying bacteria. This targeted design enhances nitrogen removal efficiency compared to traditional processes, offering improved control over operational conditions and enabling more consistent effluent quality. The separation of the two sludge populations minimizes the interference between the aerobic and anoxic microbial processes, resulting in a more robust and adaptable treatment system.

9.4 Mass Balance Analysis for Combined Predenitrification and Aerobic Processes

In this section, we develop a set of mass balance equations that form the theoretical foundation for designing and analyzing a wastewater treatment system employing both predenitrification and aerobic oxidation. These equations account for the carbon (BOD), nitrogen, and oxygen transformations occurring within the reactors and help quantify key performance parameters such as oxygen demand, biomass production, and nutrient removal. By establishing steady-state balances and defining clear notations, the following derivations provide insight into how the dual-stage process can achieve improved efficiency compared to aerobic oxidation alone.

For clarity, we use the following notation:

- Q = influent flow rate (L/d or m^3/d)
- S = soluble substrate (BOD) concentration (mg/L)
  - S_0: Influent BOD
  - S_a: BOD concentration leaving the anoxic reactor
  - S_e: BOD concentration in the final effluent

- N = nitrogen species (mg N/L)
  - $N_{TKN,in}$: Influent total Kjeldahl nitrogen (mainly as NH^{4+})
  - N_{NH_4}: Ammonium nitrogen
  - N_{NO_3}: Nitrate (and nitrite) nitrogen
- X = biomass concentration (mg/L)
- Y = yield coefficient (mg biomass produced/mg substrate consumed)
- V_A = volume of the anoxic (predenitrification) reactor
- V_{AE} = volume of the aerobic reactor

In addition, we assume that a fraction of the nitrified nitrate from the aerobic reactor is recycled to the anoxic reactor to serve as the electron acceptor for denitrification.

A. Carbon (BOD) Mass Balances

a. Anoxic (Predenitrification) Reactor

The carbon balance in the anoxic reactor accounts for the use of organic substrate (BOD) for (i) denitrification and (ii) biomass synthesis. In steady state, the input carbon equals the sum of the carbon in the effluent plus that removed by reaction:

$$QS_0 = QS_a + V_A r_{S,anox} + Y_A V_A r_{X,anox}$$

where:

$r_{S,anox}$ is the rate of substrate consumption ($mg \frac{BOD}{L} \cdot d$) by heterotrophic bacteria in the anoxic reactor, and
$r_{X,anox}$ is the rate of biomass formation.

b. Aerobic Reactor

The aerobic reactor further oxidizes the remaining BOD and nitrifies ammonium. Its carbon balance is:

$$QS_a = QS_e + V_{AE} r_{S,aer} + Y_{AE} V_{AE} r_{X,aer}$$

where:

$r_{S,aer}$ is the rate of substrate oxidation (mg BOD/L·d) in the aerobic reactor, and
Y_{AE} is the biomass yield coefficient in the aerobic reactor.

B. Nitrogen Mass Balances

The overall nitrogen balance must account for:

- **Assimilation** of nitrogen into new biomass, and
- **Conversion** of ammonium to nitrate (nitrification) followed by **denitrification** of the recycled nitrate to dinitrogen gas (N_2).

a. Overall Nitrogen Balance

For the complete system (flowing from influent to final effluent), assuming that most of the TKN is initially as NH_4^+:

$$QN_{TKN,in} = QN_{eff} + \Delta N_{assim} + N_{denit}$$

where:

N_{eff} is the nitrogen remaining in the final effluent,
ΔN_{assim} is the nitrogen incorporated into biomass, and
N_{denit} is the nitrogen removed as N_2 through denitrification.

b. Reactor-Specific Balances

Aerobic Reactor (Nitrification):

In the aerobic stage, ammonium is oxidized to nitrate according to the stoichiometric equation:

$$NH_4^+ + 2O_2 \rightarrow NO_3^- + 2H^+ + H_2O$$

The mass balance for ammonium may be written as:

$$QN_{NH_4,anox} = V_{AE} r_{NH_4,nit} + N_{NO_3,recycle}$$

where:

$r_{NH_4,nit}$ is the nitrification rate $(mg NH_4^+ - \frac{N}{L} \cdot d)$, and
$N_{NO_3,recycle}$ is the amount of nitrate recycled to the anoxic reactor.

Anoxic Reactor (Denitrification):

In the anoxic reactor, the recycled nitrate is reduced to nitrogen gas. A simplified denitrification reaction using a representative organic substrate (such as acetate) is:

$$NO_3^- + \text{organic matter} \rightarrow N_2 + CO_2 + H_2O$$

Its mass balance is:

$$N_{NO_3,recycle} = V_A r_{NO_3,denit}$$

where $r_{NO_3,denit}$ is the nitrate reduction rate (mg $NO_3^- - N/L \cdot d$).
Thus, the overall nitrogen removal by denitrification is:

$$N_{denit} = V_A r_{NO_3,denit}$$

C. Oxygen Mass Balance.

Oxygen is used in the aerobic reactor for both:

- **Oxidation of organic matter (BOD removal):**

  A typical stoichiometric factor is about 1.42 mg O_2 per mg BOD oxidized.

- **Nitrification of ammonium:**

The reaction

$$NH_4^+ + 2O_2 \rightarrow NO_3^- + 2H^+ + H_2O$$

implies approximately $2\,mg\,O_2$ are required per $mgNH_4^+ - N$ oxidized (after appropriate conversion).

However, by routing a portion of the available organic matter to the anoxic denitrification step, the process "saves" oxygen because nitrate reduction does not require O_2. It is common to quote an oxygen saving of about $2.86\,mg\,O_2$ per $mg\,NO_3^- - N$ reduced compared to the oxygen that would be needed if that organic matter were oxidized aerobically.

Thus, the net oxygen demand for the combined system is given by:

$$O_{2,total} = \left(1.42\,S_0 + 2N_{NH_4,in}\right) - \Delta O_2$$

With

$$\Delta O_2 \approx 2.86\left(N_{NO_3,recycle}\right)$$

where all terms are expressed in consistent mass units (e.g., $mg\frac{O^2}{L}$), after converting nitrogen and BOD concentrations to oxygen-equivalent values where necessary.

Practice:

An industrial wastewater is characterized as follows: $BOD_L = 5{,}000$ mg/L, TKN = 150 mg/L, $NO_3^- + NO_2 - N = 0\,mg/L$, SS = 250 mg/L, pH = 8.4, alkalinity = 2,000 mg/L, and TDS = 4,000 mg/L. Evaluate the technical and economic suitability of using a predenitrification scheme to treat this wastewater. In other words, is it possible to do predenitrification? Compared to just aerobic oxidation, what are the effects of predenitrification on O_2 usage, sludge production, and pH?

Solution:

Given Wastewater Characteristics:

- $BOD_L = 5{,}000$ mg/L → High organic loading
- TKN = 150 mg/L → High nitrogen content (likely mostly as NH^{4+})
- $NO_3^- + NO_2^- - N = 0\frac{mg}{L}$ → No oxidized nitrogen
- SS = 250 mg/L → Moderate solids

- pH = 8.4 → Slightly alkaline
- Alkalinity = 2,000 mg/L → Very high (buffering capacity present)
- TDS = 4,000 mg/L → Elevated; may affect microbial communities depending on salinity tolerance

Is predenitrification possible?

Predenitrification refers to the process where wastewater first enters an anoxic tank (no oxygen, but nitrate present), where heterotrophic bacteria reduce nitrate $\left(NO_3^-\right)$ to nitrogen gas using organic carbon (BOD) as an electron donor, followed by an aerobic stage for nitrification and remaining BOD removal.

Problem: $NO_3^- + NO_2^- = 0\frac{mg}{L}$

- There is no existing nitrate/nitrite in the influent, so predenitrification cannot begin directly, because denitrification requires nitrate.
- You must first nitrify the TKN $\left(NH_4^+\right)$ to nitrate in the aerobic stage, then recycle the nitrified mixed liquor to the anoxic zone → this becomes Modified Ludzack-Ettinger (MLE) configuration.

Thus, predenitrification is technically feasible if internal nitrate recycling is used after aerobic nitrification. The system must be designed accordingly.

Comparison: Predenitrification versus Aerobic Oxidation

(a) **Oxygen Usage**

- Aerobic oxidation only:
  - Oxygen required for BOD oxidation:$\sim 1.42\,g\frac{O_2}{g}BOD$
  - Oxygen required for nitrification:$\sim 4.57\,g\frac{O_2}{g}NH_4^+ - N$
  - Total oxygen demand = BOD + TKN oxidation

According to the reaction, 2 mol of O_2 are required to oxidize 1 mol of $NH_4^+ - N$. Therefore, the oxygen mass required per mole of $NH_4^+ - N$ is:

$$2\,\text{moles}O_2 \times \frac{32\ g}{mol} = 64\,g\ O_2$$

Since 1 mol of $NH_4^+ - N$ has a mass of about 14 g, the oxygen requirement is:

$$64\,g\,O_2/14g\ N \approx 4.57\,g\,O_2\,\text{per}\,g\,NH_4^+ - N$$

- Predenitrification:
  - Some of the BOD is consumed in anoxic denitrification → less oxygen is needed
  - Denitrification saves $\sim 2.86\,\mathrm{g}\ O_2$ per g $NO_3^- - N$ reduced (because it replaces aerobic oxidation)

Thus, predenitrification reduces overall oxygen demand and thus energy cost for aeration.

(b) **Sludge Production**

- Aerobic systems produce more biomass per unit substrate removed.
- In denitrification, biomass yield is lower than aerobic oxidation.

Thus, predenitrification produces less sludge, which reduces disposal costs.

(c) **pH Impact**

- Nitrification lowers pH (produces H^+ ions):
- $NH_4^+ \rightarrow NO_3^-$ produces 2 moles $\frac{H^+}{\mathrm{mol}}N$ → Acidifies medium
- Denitrification consumes H^+ → Alkalinity is recovered: $NO_3^- \rightarrow N_2$ consumes $\sim 3.6\,\mathrm{mg}$ as $CaCO_3$ per mg $NO_3^- - N$

Thus, predenitrification offsets acidification, helping to maintain pH, especially important since the wastewater starts at pH 8.4 and has high alkalinity (good buffering).

Technical and Economic Suitability

Parameter	Predenitrification Effect	Benefit
Nitrate Availability	Absent initially → Needs recycling	Design-dependent
BOD: N ratio	$BOD\langle sub\rangle L\langle\overline{sub}\rangle\overline{TKN} = \frac{5000}{150} \approx 33$	High—good for denitrification
Alkalinity	2000 mg/L	Supports nitrification
Oxygen demand	Reduced	Energy saving
Sludge production	Reduced	Lower disposal cost
pH	Better maintained	Operationally stable

Final recommendation

Yes, predenitrification is technically feasible and economically favorable, but only if:

- You include an aerobic nitrification step, followed by internal nitrate recycling to an anoxic zone (MLE configuration).

- The very high BOD and alkalinity levels make it especially well-suited to this approach.

It's important to note that Total Kjeldahl Nitrogen (TKN) serves as a key indicator in nitrification–denitrification processes, representing the combined organic and ammonia nitrogen available for transformation into nitrate and its subsequent reduction to nitrogen gas.

Practice: In a study of denitrification of an industrial wastewater, 120 mg/L acetate was used in the removal of 30 mg/l nitrate-nitrogen. Estimate f_s and f_e.

Solution:

First step—Convert Acetate to Electron Equivalents

Acetate used: 120 mg/L.

Molecular weight of acetate: 60 mg/mmol.

Moles of acetate per liter:

$$120\,\frac{\frac{\text{mg}}{\text{L}}}{60}\,\frac{\text{mg}}{\text{mmol}} = 2\,\frac{\text{mmol}}{\text{L}}$$

Electron yield of acetate: Each mole of acetate provides 8 electron equivalents.

$$2\,\frac{\text{mmol}}{\text{L}} \times 8 = 16\,\frac{\text{me}^{-}\text{eq}}{\text{L}}$$

Hence, there are 16 me^{-} eq/L of reducing power coming from the added acetate.

Second step—Convert Nitrate–N to Electron Demand

Nitrate–N removed: 30 mg/L.

Molecular weight of nitrogen (N): 14 mg/mmol.

Moles of nitrate–N per liter:

$$30\,\frac{\frac{\text{mg}}{\text{L}}}{14}\,\frac{\text{mg}}{\text{mmol}} \approx 2.14\,\frac{\text{mmol}}{\text{L}}$$

Electrons required to reduce 1 mole $NO_3^{-}-N$ to N_2 : 5 electron equivalents

$$2.14\,\frac{\text{mmol}}{\text{L}} \times 5 = 10.7\,\frac{\text{me}^{-}\text{eq}}{\text{L}}$$

So, 10.7 $\frac{\text{me}^{-}\text{eq}}{\text{L}}$ go toward reducing nitrate to nitrogen gas (i.e., for energy).

Final step: Partition of Electrons Between Energy and Synthesis

Total electrons (from acetate): $16\frac{me^{-}eq}{L}$.

Electrons used for nitrate reduction: $10.7\frac{me^{-}eq}{L}$.

Electrons available for cell synthesis: $16 - 10.7 = 5.3\frac{me^{-}eq}{L}$.

We define:

f_e = electrons for energy/total electrons.

f_s = electrons for synthesis/total electrons.

Thus, $f_e = \frac{10.7}{16} \approx 0.67, fs = \frac{5.3}{16} \approx 0.33$.

Practice: An industrial waste contains nitric acid (HNO_3) at 7 mM. An aspiring young environmental engineer, eager to impress her new employer, suggests they use waste syrup as the carbon source. Syrup has a BOD_L of 100,000 mg/L and is basically pure carbohydrate ($C_6H_{12}O_6$). The engineer says to operate the system with $\theta_x = 8$ d. If $f_d = 0.8, f_s^0 = 0.6$, $K_d = 0.08\ d^{-1}$, $q_{max} = 18\ g\frac{BOD_L}{gVSS_a} - d$, and $K = 20\,mg\frac{BOD_L}{L}$, how much syrup (volume) must be added to remove essentially all the nitric acid? What is the absolute minimum amount of alkalinity needed to neutralize the acid in the treated wastewater?

Solution:

1. First step: Overview of the Problem

- **Given**: A waste stream contains 7 mM of nitric acid (HNO_3). Because HNO_3 fully dissociates, we essentially have 7 mM $NO_3^{-}-N$.
- **Goal**:
    1. Find how much biodegradable organic (i.e., "syrup" with 100,000 mg BOD_L/L) must be added to remove nearly all of the nitric acid via denitrification.
    2. Determine the **minimum alkalinity** required to neutralize the acid in the treated wastewater.

Key microbial/kinetic parameters are provided to handle yield effects and cell synthesis (rather than using a simplified $\frac{COD}{NO_3}$ ratio). Hence, the final required BOD is calculated from a more rigorous stoichiometric approach, including the fraction of electrons going to new cell growth (f_s) and that going to energy (f_e).

2. Second step: Stoichiometric Basis

$$NO_3^{-} + \text{organic substrate} \rightarrow N_2 + \text{cells (some fraction)} + \text{other products}$$

In particular, for an operating sludge age of$\theta_x = 8d$, plus given values of$f_d, f_s^0 = 0.6$, K_d, q_{max} etc., a stoichiometric reaction can be derived to find the net BOD consumed per mole of $NO_3^-{-}N$ reduced. This includes:

- The fraction of electrons for synthesis (f_s) vs. for energy (f_e).
- Cell decay and net yield considerations at $\theta_x = 8d$.
- Residual BOD after accounting for biomass growth.

From given value:

$$f_s = \frac{f_s^0\left[1 + (1{-}f_d) \times K_d \times \theta_x\right]}{(1 + K_d \times \theta_x)}$$

$$f_s = \frac{0.6\frac{e^- \text{ cells}}{e^- \text{ total}}\left[1 + (1{-}0.8) \times 0.08\,\text{days}^{-1} \times 8\,\text{days}\right]}{\left(1 + 0.08\,\text{days}^{-1} \times 8\,\text{days}\right)} = 0.41\frac{e^- \text{ cells}}{e^- \text{ total}}$$

Thus, $f_e = 1{-}f_s = 1{-}0.41 = 0.59\frac{e^- \text{ acceptor}}{e^- \text{ total}} f_e = 1{-}f_s = 1{-}0.41 = 0.59\frac{e^- \text{ acceptor}}{e^- \text{ total}}$

Now, you need to write biological reaction for glucose as donor, nitrate as electron acceptor and cell synthesis reaction with nitrate as nitrogen source.

$$-R_d : \frac{1}{24}C_6H_{12}O_6 + \frac{1}{4}H_2O = \frac{1}{4}CO_2 + H^+ + e^-$$

$$R_c : \frac{1}{28}NO^{3-} + \frac{5}{28}CO_2 + \frac{29}{28}H^+ + e^- = \frac{1}{28}C_5H_7O_2N + \frac{11}{28}H_2O$$

$$R_a : \frac{1}{5}NO^{3-} + \frac{6}{5}H^+ + e^- = \frac{1}{10}N_2 + \frac{3}{5}H_2O$$

From above reactions:

$f_sR_c : 0.41R_c$

$0.0146NO^{3-} + 0.073CO_2 + 0.424H^+ + 0.41e^- = 0.0146C_5H_7O_2N + 0.161H_2O$

$f_eR_a : 0.59R_a$

$0.118NO^{3-} + 0.71H^+ + 0.59e^- = 0.059N_2 + 0.354H_2O$

$-R_d : 0.042C_6H_{12}O_6 + 0.25\ H_2O = 0.25CO_2 + H^+ + e^-$

Now add: $f_sR_c + f_eR_a + (-R_d)$ for complete biological reaction.

$$\begin{aligned} R : &\ 0.04\ C_6H_{12}O_6 + 0.133NO^{3-} + 0.13H^+ \\ &= 0.0146C_5H_7O_2N + 0.06N_2 + 0.177CO_2 + 0.27H_2O \end{aligned}$$

From here:

BOD required for each mole of $NO_3^-{-}N$ removed = 0.04 × 180 (molecular weight of glucose)/0.133 × 14 = 3.86 g BOD/g N.

Multiplying that by the total moles of $NO_3^- - N$ present (7 mM) in 1 L (= 98 mg N/L) completes the stoichiometric calculation.

Thus, total BOD requirement ≈ 379 mg BOD/L after combining all factors.

3. Third step: Volume of Syrup Needed

- **Syrup strength:** 100,000 mg BOD_L/L of syrup.
- **BOD needed:** 379 mg BOD/L of wastewater.

Hence, the **volume** of syrup per liter of wastewater is:

Volume of syrup per L wastewater =

$$V = 379\,\frac{\text{mg BOD}}{\text{L wastewater}} \times \left(\frac{1\,\text{L syrup}}{100,000\,\text{mg BOD}}\right)$$
$$= 0.00379L \approx 3.8\,\text{mL of syrup per L of wastewater}$$

4. Minimum Alkalinity Required

In principle, each NO_3^- (or HNO_3) introduces 1 equivalent of acidity, so removing 7 mM $NO_3^-{-}N$ implies neutralizing 7 meq of acid per liter. However, as per biological equation, additional CO_2 formation can create further acidity.

Calculating from biological equation:

$$\left(\frac{0.177\,\text{mole}\,O_2}{0.133\,\text{mole}\,NO_3^-}\right) \times \left(\frac{7\text{mM}\frac{NO_3^-}{\text{L}}}{1\frac{\text{eq}}{\text{mole}}CO_3^{2-}}\right)$$
$$= 9.31\,\text{mM alkalinity}$$

Alkalinity is expressed as equivalent weight of $CaCO_3$.

Thus, the final alkalinity = 9.31 × 50 = 465.7 mg $\frac{CaCO_3}{L}$.

5. Final Answers

1. **Syrup Volume**:

3.8 mL of syrup per L of wastewater

2. **Minimum Alkalinity (as** $CaCO_3$):
   465.7 mg $\frac{CaCO_3}{L}$

Practice: What concentration of lactate would be required for denitrification of 60 mg/L nitrate-nitrogen in a CSTR with settling and recycle in which θ_x was maintained at 6 d and f_s= 0.6?

Solution:

Step 1. Determine the Theoretical Electron Demand for Nitrate Reduction

We are given that the wastewater contains 60 mg/L nitrate-nitrogen. Since nitrate ($NO_3^- - N$) is reduced to N_2, the typical reduction requires about 5 electrons per nitrogen atom (Reaction #7, Table 2.2).

1. Convert nitrate–N to moles per liter:

$$\text{Moles of N} = 60\,\frac{\text{mg}}{\text{L}} - 14\,\frac{\text{mg}}{\text{mmol}} = 4.29\,\frac{\text{mmol N}}{\text{L}}$$

2. Calculate the total electron equivalents required:

$$\text{Total electrons(theoretical)} = 4.29\,\frac{\text{mmol N}}{\text{L}} \times 5\,\frac{\text{mol e}^-}{\text{mol N}} = 21.45\,\frac{\text{mmol e}^-}{\text{L}}$$

Step 2. Account for Electron Partitioning in the Denitrification Process

In practice, not all electrons from the organic substrate go toward reducing nitrate; a fraction is used for cell synthesis.

Here the problem states $f_s = 0.6$, so approximately 60% of the electrons are used for biomass synthesis and only the remaining 40% (i.e. $f_e = 1 - f_s = 0.4$) are available for energy (i.e. for denitrification).

Thus, to supply 21.45 $\text{mmol}\frac{\text{e}^-}{\text{L}}$ for nitrate reduction (using only 40% of the substrate's electrons), the total electron delivery that must be provided by lactate is:

$$\text{Total electrons required from lactate} = \frac{\left(21.45\,\frac{\text{mmol e}^-}{\text{L}}\right)}{0.4} = 53.63\,\frac{\text{mmol e}^-}{\text{L}}$$

Step 3. Determine the Lactate Requirement Using Electron Equivalence

Lactate (assumed as $C_3H_5O_3^-$) is essentially a carbohydrate. One mole of lactate supplies about 12 electron equivalents (Reaction #11, Table 2.3).

1. Calculate the moles of lactate needed:

$$\text{Moles lactate} = \frac{\left(53.63\,\frac{\text{mmol e}^-}{\text{L}}\right)}{\left(12\,\frac{\text{mmol e}^-\text{mmol}}{\text{lactate}}\right)} = 4.47\,\frac{\text{mmol lactate}}{\text{L}}$$

2. Convert this to a mass concentration. The molecular weight of lactate is about 90 g/mol,

$$\text{Mass of lactate} = 4.47 \times 10^{-3}\,\frac{\text{mol}}{\text{L}} \times 90\,\frac{\text{g}}{\text{mol}} = 0.402\,\frac{\text{g}}{\text{L}}$$

Final Answer

Approximately 0.40 g of lactate per liter of wastewater would be required to denitrify 60 mg/L nitrate–nitrogen in a CSTR (with settling and recycle) operating at an SRT (θ_x) of 6 days.

9.5 Biological Phosphorus Removal Process

Biological phosphorus removal (BPR) is a process in which specific microorganisms remove phosphorus from wastewater through natural metabolic pathways, rather than relying on chemical precipitation methods. In this process, microorganisms assimilate soluble phosphate and store it intracellularly in the form of polyphosphate granules. This biologically mediated uptake of phosphorus can occur under various environmental conditions and is observed in both natural ecosystems and engineered wastewater treatment systems.

BPR serves as a general term encompassing all microbial mechanisms for phosphate removal. However, the efficiency and reliability of phosphorus removal in early BPR systems were often limited due to inconsistent environmental conditions and the lack of controlled microbial selection. These limitations led to the development of more structured and optimized approaches, giving rise to what is now known as **Enhanced Biological Phosphorus Removal (EBPR)**.

Enhanced biological phosphorus removal (EBPR) is a process that uses bacteria to remove phosphorus from wastewater. It is a more efficient and environmentally friendly way to remove phosphorus than chemical precipitation. This process depends on a specialized group of microorganisms known as **polyphosphate-accumulating organisms (PAOs)**. These bacteria have the unique ability to uptake and store phosphate from wastewater in quantities that exceed their metabolic growth requirements. This phosphate uptake occurs under a specifically designed regime of alternating anaerobic and aerobic or anoxic conditions, which creates a selective advantage for PAOs in the microbial community (Tchobanoglous et al. 2003).

In early microbiological investigations of biological phosphorus removal, bacteria belonging to the genus *Acinetobacter* were initially identified as the primary organisms responsible for phosphate accumulation in activated sludge systems (Boll 1988). These gram-negative, widely distributed bacteria were considered essential aerobic heterotrophs, though subsequent research has also demonstrated their capacity to participate in nitrogen metabolism under certain conditions.

Acinetobacter species preferentially utilize short-chain organic acids as carbon sources and are capable of storing excess phosphorus intracellularly as polyphosphate granules. Additionally, they can accumulate carbon reserves in the form of poly-β-hydroxybutyric acid (PHB), forming visible intracellular storage granules. This dual storage capability made them a key subject of early enhanced biological phosphorus removal (EBPR) studies.

A foundational contribution to the understanding of their role came from Wentzel et al. (1985), who developed one of the first mathematical models for Bio-P removal. Their work, based on both natural and enriched cultures of *Acinetobacter*, laid the groundwork for the calculation of biological phosphorus removal rates and remains influential in the field to this day.

Further, among the PAOs, the most widely studied and well-characterized organism is 'Candidatus *Accumulibacter phosphatis*', commonly referred to as *Accumulibacter*. This organism belongs to subclass 2 of the Betaproteobacteria and is phylogenetically related to the genus *Rhodocyclus* (Hesselmann and Werlen 1999). *Accumulibacter* plays a central role in EBPR systems by cycling polyphosphate granules within its cells as a means of energy storage and release, enabling efficient phosphate removal from wastewater.

In addition to *Accumulibacter*, research has identified other microbial taxa capable of functioning as PAOs. Notably, certain gram-positive Actinobacteria, particularly those related to the genus *Tetrasphaera*, have also been implicated in phosphorus accumulation (Nguyen et al. 2011). While *Tetrasphaera* species exhibit different metabolic pathways compared to *Accumulibacter*—often involving fermentation and amino acid uptake under anaerobic conditions—they are now considered important contributors to phosphorus removal in full-scale wastewater treatment plants.

The co-existence and ecological interaction of diverse PAO populations, including both *Accumulibacter* and *Tetrasphaera*-related organisms, are critical to the stability and performance of EBPR processes. Ongoing research using molecular tools such as fluorescence in situ hybridization (FISH) and metagenomics continues to refine our understanding of these functional microbial communities and their roles in nutrient removal.

BPR is a two-step process:

A. **Phosphorus Release (Anaerobic Stage)**

- **Mechanism**: Under anaerobic conditions, PAOs utilize the energy stored in the intracellular polyphosphate granules by breaking the polyphosphate bonds. This energy is used to take up readily biodegradable compounds, primarily short-chain volatile fatty acids (VFAs).
- **Storage of Carbon**: The VFAs are then polymerized and stored within the cells as poly-β-hydroxy-alkanoates (PHAs), with poly-β-hydroxy-butyrate (PHB) being the most common product. PHB is synthesized by bacteria from 2 mol of acetyl coA (recall glycolysis and pyruvate oxidation and TCA cycle in Chap. 4).
- **Importance of Anaerobic Conditions**: The absence of alternative electron acceptors like oxygen (O_2) or nitrate (NO_3^-) is crucial during this phase, as their presence would otherwise inhibit the storage process.

B. **Phosphorus Uptake (Aerobic or Anoxic Stage)**

- **Mechanism**: Following the anaerobic phase, the system is transitioned into an aerobic (or sometimes anoxic) phase. In this environment, PAOs metabolize the

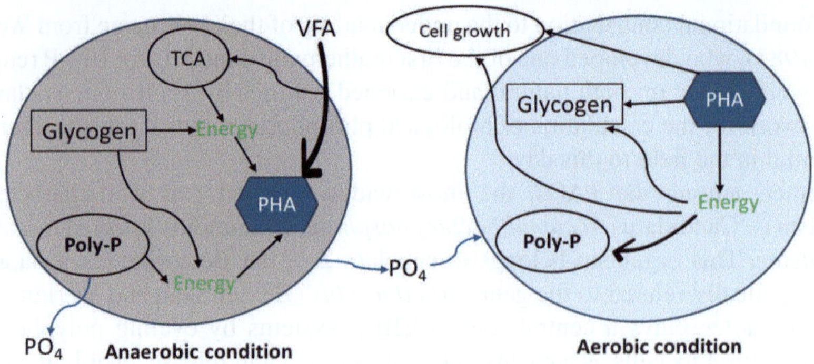

Fig. 9.3 A conceptual representation of BPR theory

stored PHAs to generate energy required for growth, maintenance, and the active uptake of orthophosphate from the wastewater.

- **Formation of Polyphosphate**: As the PAOs use the energy from PHAs, they take up phosphorus in amounts that exceed their immediate metabolic requirements. The excess phosphorus is stored intracellularly as polyphosphate, resulting in the net removal of phosphorus from the wastewater.

A conceptual representation of BPR theory is presented in Fig. 9.3.

The key steps for achieving BPR at wastewater treatment plants are illustrated in the simplified diagram in Fig. 9.4, where the step (1) is Return activated sludge from the secondary clarifier, which contains PAOs, is added to the influent wastewater; (2) is anaerobic zone, where PAOs break polyphosphate bonds to generate energy; (3) is the aerobic zone, where the PAOs use oxygen to metabolize the stored PHAs to generate energy for growth and maintenance; and (4) is clarifier, where PAOs settle to the bottom with the rest of the activated sludge and phosphorus stored in the PAOs is removed with the waste activated sludge.

BPR can be implemented in a variety of wastewater treatment systems, including:

Activated sludge systems: BPR can be implemented in activated sludge systems by adding an anaerobic zone to the system. The anaerobic zone creates an environment where PAOs can grow and store phosphorus.

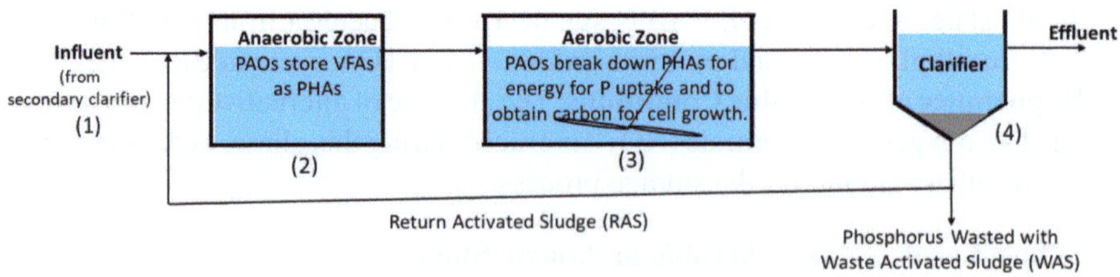

Fig. 9.4 Biological phosphorus removal in wastewater treatment plant

Membrane bioreactors: BPR can also be implemented in membrane bioreactors. Membrane bioreactors use membranes to separate solids from the wastewater. This allows the PAOs to grow and store phosphorus in the bioreactor without being removed by the membranes.

BPR is a reliable and effective method for removing phosphorus from wastewater. It is a valuable tool for wastewater treatment plants that are looking to improve the quality of their effluent. However, it is important to note that chemical precipitation is the most common method used to remove phosphorus from wastewater, typically involving the addition of salt or other precipitating agents. This technique is often employed simultaneously with biological phosphorus removal (BPR) to reduce both the cost of chemical dosing and sludge production. Phosphorus precipitation can be implemented during the primary or secondary clarification stages, or even at advanced/tertiary filtration stages of wastewater treatment. Depending on the process design, chemicals can be applied as direct precipitation, pre-precipitation, simultaneous precipitation, or post-precipitation. Common chemicals used for this purpose include metal salts and lime.

9.6 PhoStrip Process

Traditional biological phosphorus removal (BPR) process integrated into the main flow often suffer from unstable phosphorus release and uptake rates, tight process control requirements, and high sludge production indices. The PhoStrip process overcomes these limitations by relocating the anaerobic phosphorus-release step into a side-stream sludge recycle loop, decoupling it from the main hydraulic flow and influent variations. The PhoStrip process is a biological phosphorus removal technology that builds on the principles of the activated sludge system while specifically targeting phosphorus reduction. As illustrated in Fig. 9.5, it is designed to promote the proliferation of phosphorus-accumulating organisms (PAOs) under controlled environmental conditions. This process finds its niche among enhanced biological phosphorus removal (EBPR) strategies and offers a cost-effective and sustainable alternative to chemical precipitation methods.

9.6.1 Key Advantages of the Sidestream Approach

By operating entirely within the sludge recirculation circuit, PhoStrip delivers several performance benefits:

- **Independence from Influent Fluctuations**: Treatment is unaffected by swings in flow rate, temperature, or influent COD:N:P ratios.

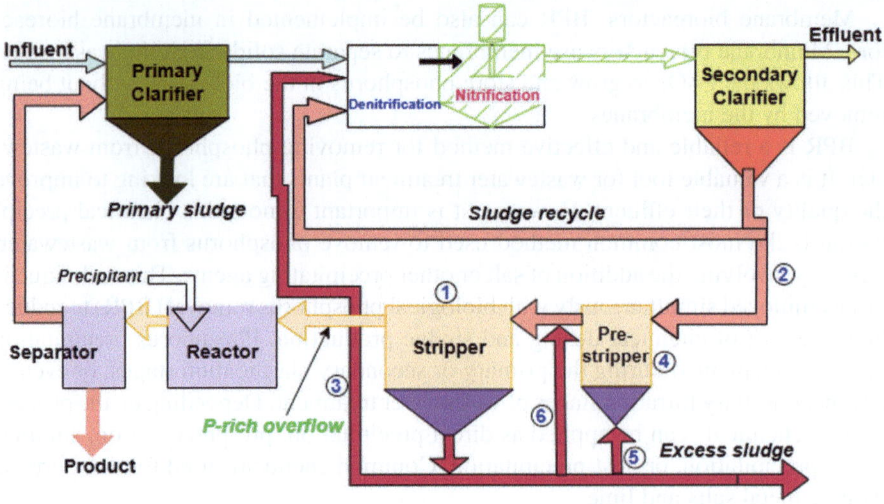

Fig. 9.5 An overview of PhoStrip side stream process. Adapted from Kaschka and Weyrer (1999)

- **Enhanced Biomass Concentration**: The mixed liquor in the side-stream can be up to twice as concentrated as in the mainstream, giving phosphorus-releasing organisms a strong selection advantage.
- **Increased Anaerobic Volumetric Loading**: Higher MLSS in the sidestream allows more short-chain volatile fatty acids (VFAs) to be produced per unit volume under anaerobic conditions.
- **Decoupled Contact Times**: Microbial contact time in the sidestream reactor can be extended up to tenfold compared to a mainstream anaerobic zone, despite a much smaller tank volume.

9.6.2 *The Stripper Tank*

At the heart of PhoStrip is the **Anaerobic Stripper Tank**, which functions similarly to a sludge thickener but is operated under strict anaerobic conditions. Within this unit:

1. **Phase Separation and Extended Contact**: Sludge and supernatant separate by sedimentation, allowing microorganisms to remain in contact with their own soluble products far longer than the actual hydraulic detention time.
2. **Sludge Thickening**: Solids concentration increases by up to 4%, boosting VFA generation rates.
3. **Long Anaerobic Retention**: Sludge can be stored anaerobically for up to 20 h, further enhancing VFA production.

4. **Particulate Hydrolysis**: Particulate COD and waste biomass are hydrolyzed into readily fermentable, short-chain organic acids—the essential electron donors for phosphorus release.
5. **Phosphate Release**: In the absence of external electron acceptors, polyphosphate-accumulating organisms degrade intracellular polyphosphate stores, releasing orthophosphate into the sidestream.
6. **Selection and Partial Bypass**: Because of the elevated MLSS and extended contact, only 20–45% of the sludge recycle needs sidestream treatment—dramatically reducing the required anaerobic tank volume.

9.6.3 Ancillary Sidestream Units

- **PreStripper (Denitrification Reactor)**: To prevent nitrate in the recycle from inhibiting phosphorus release, a short-contact anoxic reactor can reduce NO_3^- to < 2 mg N/L. VFAs from the main stripper are diverted here as the sole carbon source.
- **Internal Recirculation**: A portion of the stripped sludge is recirculated within the stripper tank to wash out released phosphate and redistribute VFAs throughout the reactor volume. This "back-wash" loop can also feed the PreStripper when combined with VFA sidestream dosing.

The PhoStrip process consists of three stages:

Aerobic Stage: The initial aerobic stage serves a dual purpose. First, it supports the growth of conventional activated sludge bacteria that metabolize organic matter, thereby reducing the biochemical oxygen demand (BOD) of the influent wastewater. Second, it helps maintain a baseline population of microorganisms before the selective enrichment of PAOs occurs.

In this stage, oxygen is abundant, facilitating rapid aerobic degradation of organic carbon and fostering the growth of biomass. The organic material is oxidized to produce carbon dioxide and energy for microbial growth.

Anoxic Stage: The anoxic (or more precisely, anaerobic) stage is crucial for the enrichment and activity of PAOs. Under oxygen-limited conditions, PAOs initiate a unique metabolic sequence that leads to phosphorus release followed by uptake.

During this phase, PAOs utilize the available organic substrates (such as volatile fatty acids) to synthesize intracellular storage compounds like poly-β-hydroxyalkanoates (PHAs). This is coupled with the hydrolysis of intracellular polyphosphate reserves, releasing phosphate into the liquid phase initially. The biochemical "reset" during this stage prepares the PAOs for enhanced phosphorus uptake in the subsequent aerobic phase.

Aerobic Polishing Stage: The final aerobic stage is employed to "polish" the wastewater by ensuring complete oxidation of residual organic compounds and re-establishing aerobic conditions favorable for phosphorus uptake.

When the wastewater is returned to an aerobic environment, PAOs utilize the energy derived from the oxidation of stored PHAs to take up phosphate from the wastewater in excess of their immediate metabolic needs. This phosphorus is converted into intracellular polyphosphate, effectively reducing the dissolved phosphorus concentration in the effluent.

The efficiency of the PhoStrip process hinges on the metabolic versatility of PAOs. Their ability to rapidly switch between anaerobic and aerobic metabolisms allows them to accumulate phosphorus, making them highly effective in lowering overall phosphorus levels. However, this same capacity means the process must be carefully managed to avoid proliferation of competing microorganisms, such as glycogen-accumulating organisms (GAOs), which can interfere with phosphorus uptake.

The PhoStrip process integrates a series of well-defined aerobic and anaerobic phases to selectively enrich PAOs, leading to efficient biological phosphorus removal. By leveraging the unique metabolic behavior of PAOs, the process not only minimizes the phosphorus load in the treated effluent but also offers environmental and economic benefits compared to traditional chemical treatment methods. With careful monitoring and operational control, the PhoStrip process stands as a promising technology for sustainable wastewater treatment.

Practice: You have a wastewater with the following characteristics:

$$Q = 2,500\frac{m^3}{d};\ BOD = 450\frac{mg}{L};\ TKN = 50\frac{mg}{L};$$
$$PO_4^{3-} - P = 12\frac{mg}{L};\ and SS = 300\frac{mg}{L}$$

You wish to treat this wastewater directly by a Pho-Strip process for enhanced biological phosphorus removal. If you maintain an SRT of 4 d, can you meet a 1 mg/L standard for phosphate-P? Why is the design feasible or infeasible? You may assume that $q_{max} = 12\,mg\frac{BOD}{mg}VSS - d$, $K = 10\,mg\frac{BOD}{L}$, $K_d = \frac{0.15}{d}$, $f_d = 0.8$, and $Y = 0.45\,mg\frac{VSS}{mg}BOD$.

Solution:

Step 1: Required phosphorus removal.

- Influent $PO_4^{3-} - P = 12$ mg/L
- Effluent target = 1 mg/L
- $\Delta P = 11\,mg\,P\,per\,liter \Rightarrow$ at $Q = \frac{2500\,m^3}{d}$ this is $\frac{11\,mg}{L} \times 2.5 \times \frac{10^6\,L}{d} = \frac{27.5 kgP}{day}$

Step 2: P removal by ordinary biomass assimilation.

Only the phosphorus built into new heterotrophic biomass:

1. **BOD removed/day** = Influent BOD × Q = $\frac{450\,\text{mg}}{\text{L}} \times 2.5 \times \frac{10^6\,\text{L}}{\text{d}} = 1.125 \times \frac{10^9\,\text{mg BOD}}{\text{d}}$.
2. **Net biomass yield**

$$q = q_{\max}\frac{S}{K+S} = \frac{(12 \times 450)}{(10+450)} = 11.74\frac{\text{mg BOD}}{\text{mg VSS}}$$

$$\begin{aligned} Y_{net} &= Y\left(\frac{q}{q+K_d}\right) \\ &= 0.45\left(\frac{11.74}{11.74+0.15}\right) = 0.444\frac{\text{mg VSS}}{\text{mg BOD}} \end{aligned}$$

$$\begin{aligned} \text{New VSS} &= 1.125 \times \frac{10^9\,\text{mg BOD}}{\text{d}} \times \frac{0.444\,\text{mg VSS}}{\text{mg BOD}} \\ &= 5.00 \times \frac{10^8\,\text{mg VSS}}{\text{d}} = \frac{500\,\text{kg VSS}}{\text{d}} \end{aligned}$$

3. **Typical P content of biomass** ~1% of VSS ⇒ 0.01 mg P Per mg VSS

$$\text{P fixed} = \frac{500\,\text{kg VSS}}{\text{d}} \times \frac{0.01\,\text{kg P}}{\text{kg VSS}} = \frac{5\,\text{kg P}}{\text{d}}$$

$$\text{Spread over}\frac{2500\,\text{m}^3}{\text{d}} \Rightarrow \frac{5000\,\text{g P}}{2.5} \times 10^6\text{L} = \frac{2\,\text{mg P}}{\text{L}}$$

It means assimilation alone gives only ~ 2 mg P/L removal.

Step 3: P removal by luxury uptake (EBPR) fed solely by sidestream VFA.

Pho-Strip sidestream "ferments" wasted RAS to VFA, which then drives PAO uptake in the mainstream. Let's estimate how much VFA you can get:

1. **Wasted VSS available** = 500 kg VSS/d (from above).
2. **VFA yield** from fermenting mixed liquor solids = 50% of VSS ⇒ 250 kg VFA (as COD) per day.
3. **Mix this into the 2 500 m³/d influent** ⇒ $\frac{250{,}000\,\text{g COD}}{2.5} \times 10^6\text{L} = \frac{100\,\text{mg COD}}{\text{L}}$ (as VFA)
4. **EBPR stoichiometry** = 20 mg COD per mg P removed ⇒ $\frac{100\,\text{mg COD}}{\text{L}} \div 20\,\text{mg COD per mg P} = \frac{5\,\text{mg P}}{\text{L}}$ additional removal

Step 4: Total achievable P removal.

- Assimilation: 2 mg P/L
- Luxury uptake (side-stream VFA): 5 mg P/L
- Total = 7 mg P/L removed

Effluent $= 12{-}7 = \frac{5\,\text{mg P}}{\text{L}}$, which is well above the 1 mg P/L target.

Step 5: Why it fails and what's needed.

- No primary sludge is fermented in a straight Pho-Strip; you only get VFA from wasted RAS, which is limited.
- That sidestream VFA supports only 5 mg P/L luxury uptake.
- Together with assimilation, you only hit 7 mg P/L removal.

To push effluent below 1 mg P/L you must either:

1. Ferment additional solids (e.g., primary sludge) to boost VFA production, or
2. Add external carbon (acetate, methanol, etc. as precipitant), or
3. Increase sidestream recirculation drastically (but that has practical limits), or
4. Combine with chemical P-precipitation for polishing.

Final conclusion: With an SRT of 4 d and only RAS sidestream fermenter, you cannot meet a $\frac{1\,\text{mg}}{\text{L}}PO_4{-}P$ standard (you'll end up around 5 mg/L). The design is infeasible unless you augment VFA supply or add a polishing step.

Further Readings

Boll R (1988) Zur erhöhten biologischen Phosphorentfernung mit dem Belebungs-verfahren. Veröffentlichung des Instituts für Siedlungswasserwirtschaft TU Braunschweig. Heft 46:1–195

Hesselmann RPX, Werlen C, Hahn D, van der Meer JR, Zehnder AJB (1999) Enrichment, phylogenetic analysis and detection of a bacterium that performs enhanced biological phosphate removal in activated sludge. Syst Appl Microbiol 1999, 22:454–465

Cadmus Group, Inc. (2010). Nutrient Control Design Manual (STREAMS Task Order 68). Cadmus Group, Inc

Kaschka E, Weyrer S (1999). Biological elimination of Phosphorus from domestic sewage by applying the enhanced Phostrip process. In: Phostrip handbok, 4th edn. Phostrip Abwassertechnik GmbH.

Nguyen HTT, Le VQ, Hansen AA, Nielsen JL, Nielsen PH (2011) High diversity and abundance of putative polyphosphateaccumulating Tetrasphaera-related bacteria in activated sludge systems. FEMS Microbiol Ecol 76:256–267

Rittmann BE, McCarty PL (2001) Environmental biotechnology: principles and applications. McGraw-Hill Education, New York, Chicago, San Francisco, Athens, London, Madrid, Mexico City, Milan, New Delhi, Singapore, Sydney, Toronto

Tchobanoglous G, Burton F, Stensel H, Metcalf and Eddy Inc (2003) Wastewater engineering, treatment and reuse. McGraw-Hill, New York, NY (USA)

Wentzel MC, Dold PL, Ekama GA, Marais GVR (1985) Kinetics of biological phosphorous release. Wate Sci Technol 17:57–71

Test Your Understanding: MCQs

1. What is the primary role of microorganisms in environmental engineering, particularly in water and wastewater treatment?
   (A) To increase the pH of wastewater
   (B) To transform and stabilize pollutants biologically
   (C) To reduce the need for chemical treatments entirely
   (D) To increase the turbidity of water

2. Which of the following is a key difference between prokaryotes and eukaryotes?
   (A) Prokaryotes have a membrane-bound nucleus, while eukaryotes do not.
   (B) Prokaryotes lack membrane-bound organelles, while eukaryotes have them.
   (C) Eukaryotes are always unicellular, while prokaryotes are multicellular.
   (D) Prokaryotes are only found in extreme environments.

3. Why are bacteria considered ubiquitous in environmental engineering applications?
   (A) They are found only in wastewater treatment plants.
   (B) They are present in almost every environment and play crucial roles in biodegradation.
   (C) They are harmful and must be eliminated from all water sources.
   (D) They cannot survive in extreme conditions.

4. Which of the following is NOT a common way to categorize bacteria?
   (A) Shape (cocci, bacilli, spirilla)
   (B) Oxygen requirements (aerobic/anaerobic)
   (C) Color under a microscope
   (D) Gram staining (Gram-positive/Gram-negative)

V. Achal, *Notes on Biological Processes in Environmental Engineering*,
https://doi.org/10.1007/978-981-95-1610-0

5. In wastewater treatment, which group of bacteria is primarily responsible for breaking down organic matter?
   (A) Photosynthetic bacteria
   (B) Nitrifying bacteria
   (C) Heterotrophic bacteria
   (D) Methanogenic archaea

6. Which of the following best describes the importance of prokaryotes in environmental engineering?
   (A) They are only useful in laboratory research.
   (B) They play a minor role compared to chemical processes.
   (C) They are essential for nutrient cycling, pollutant degradation, and bioremediation.
   (D) They are primarily used for producing antibiotics.

7. What is the primary purpose of constructing balanced mass equations in biological wastewater treatment?
   (A) To determine the exact number of bacteria in a sample
   (B) To calculate chemical, nutrient, and energy requirements for bacterial growth
   (C) To measure the color changes in wastewater
   (D) To eliminate the need for bacterial metabolism

8. Which of the following is NOT typically included in a stoichiometric mass balance for bacterial growth?
   (A) Electron acceptors (e.g., oxygen)
   (B) Nutrients (e.g., nitrogen, phosphorus)
   (C) Bacterial genetic mutations
   (D) pH-adjusting chemicals (e.g., sulfuric acid)

9. In bacterial stoichiometry, what role do bacteria play in chemical reactions?
   (A) They are inert spectators.
   (B) They act as catalysts and also as reaction products.
   (C) They only consume oxygen without participating in reactions.
   (D) They destabilize the reaction equilibrium.

10. What is the significance of the empirical formula of bacterial cells (say, $C_5H_7O_2N$) in environmental engineering?
    (A) It helps in identifying bacterial species under a microscope.
    (B) It is used to balance stoichiometric equations for biomass growth.
    (C) It determines the color of bacterial colonies.
    (D) It predicts bacterial motility in wastewater.

11. Which term describes the fraction of substrate converted into new bacterial cells (biomass)?

    (A) Gibbs free energy
    (B) Cellular yield (Y)
    (C) Electron equivalence
    (D) Thermodynamic efficiency

12. In bacterial energetics, what is the primary use of energy released from redox reactions?

    (A) To produce light in wastewater
    (B) For cell synthesis and maintenance
    (C) To increase wastewater temperature
    (D) To neutralize pH without chemicals

13. What does a negative Gibbs free energy change (ΔG°) indicate about a reaction?

    (A) The reaction is non-spontaneous.
    (B) The reaction is at equilibrium.
    (C) The reaction is spontaneous and releases energy.
    (D) The reaction requires external catalysis.

14. Which of the following best describes the Thermodynamic Electron Equivalents Model (TEEM)?

    (A) A method to count bacterial cells microscopically
    (B) A framework to balance microbial-metabolic reactions using energy principles
    (C) A technique to measure bacterial motility
    (D) A model to predict bacterial colony colors

15. Why is maintaining proper nutrient balance (N and P) critical in biological wastewater treatment?

    (A) To ensure bacterial growth and efficient pollutant degradation
    (B) To increase the turbidity of treated water
    (C) To replace the need for electron acceptors
    (D) To eliminate all bacterial activity

16. What is the role of electron acceptors (such as oxygen, nitrate) in bacterial metabolism?

    (A) They serve as direct nutrient sources for biomass.
    (B) They accept electrons during oxidation of substrates, releasing energy.
    (C) They inhibit bacterial growth to reduce sludge production.
    (D) They are only needed for photosynthetic bacteria.

17. What does Biochemical Oxygen Demand (BOD) measure in water?

    (A) The total amount of organic and inorganic chemicals
    (B) The amount of oxygen consumed by microorganisms to degrade organic matter
    (C) The concentration of dissolved oxygen at saturation
    (D) The amount of oxygen produced by photosynthesis

18. Which of the following best describes the key difference between BOD and COD?

    (A) BOD measures only inorganic matter, while COD measures organic matter.
    (B) BOD is a biological process, while COD is a chemical oxidation process.
    (C) COD measures oxygen demand over 5 days, while BOD provides instantaneous results.
    (D) BOD is always higher than COD for the same sample.

19. Why is the BOD test typically conducted over 5 days (BOD_5)?

    (A) To match the standard incubation time for most wastewater bacteria
    (B) Because all organic matter degrades completely within 5 days
    (C) To avoid interference from photosynthetic oxygen production
    (D) It is an arbitrary time frame with no scientific basis

20. What does Theoretical Oxygen Demand (ThOD) represent?

    (A) The oxygen required to chemically oxidize a compound completely to CO_2 and H_2O
    (B) The oxygen produced during algal blooms
    (C) The residual oxygen left after BOD testing
    (D) The oxygen demand measured in anaerobic conditions

21. Which oxygen demand metric would be most useful for assessing short-term pollution impacts in a river?

    (A) BOD_5 (5-day BOD)
    (B) COD (Chemical Oxygen Demand)
    (C) ThOD (Theoretical Oxygen Demand)
    (D) DO (Dissolved Oxygen)

22. In a wastewater sample, if COD is significantly higher than BOD, what does this likely indicate?

    (A) The sample contains easily biodegradable organic matter.
    (B) The sample contains recalcitrant (hard-to-degrade) organic compounds.
    (C) The BOD test was performed incorrectly.
    (D) Photosynthetic organisms are increasing oxygen levels.

23. Which of the following statements about ThOD is FALSE?

    (A) ThOD assumes complete oxidation of a compound.
    (B) ThOD is always lower than BOD for the same substance.
    (C) ThOD can be calculated using stoichiometric equations.
    (D) ThOD does not account for microbial metabolism limitations.

24. What is the primary purpose of catabolic pathways in bacterial cells?

    (A) To synthesize complex macromolecules like DNA and proteins
    (B) To break down substrates and generate ATP for cellular energy
    (C) To store excess nutrients as glycogen or lipids
    (D) To eliminate waste products through passive diffusion

25. In glycolysis, what is the net ATP yield per glucose molecule?

    (A) 1 ATP
    (B) 2 ATP
    (C) 4 ATP
    (D) 36 ATP

26. Which of the following correctly describes the electron flow in the mitochondrial electron transport chain (ETC)?

    (A) NADH → Complex II → Complex III → Complex IV
    (B) $FADH_2$ → Complex I → Complex III → Complex IV
    (C) NADH → Complex I → Complex III → Complex IV
    (D) $FADH_2$ → Complex I → Complex II → Complex IV

27. Which of the following is NOT a product of the Krebs cycle?

    (A) NADH
    (B) $FADH_2$
    (C) ATP (or GTP)
    (D) Oxygen (O_2)

28. How many NADH molecules are generated from one glucose molecule during aerobic respiration (through glycolysis, pyruvate oxidation, and the Krebs cycle)?

    (A) 2
    (B) 6
    (C) 10
    (D) 12

29. What role does oxidative phosphorylation play in ATP production?
    (A) It directly synthesizes ATP during glycolysis.
    (B) It uses the proton gradient created by electron transport to drive ATP synthase.
    (C) It breaks down glucose into pyruvate.
    (D) It regenerates NAD^+ for fermentation.

30. Which of the following pathways generates ATP without requiring oxygen?
    (A) Krebs cycle
    (B) Oxidative phosphorylation
    (C) Glycolysis
    (D) Pyruvate oxidation

31. What is the theoretical maximum ATP yield from one glucose molecule in prokaryotes?
    (A) 2 ATP
    (B) 16 ATP
    (C) 32 ATP
    (D) 38 ATP

32. Why is the Krebs cycle considered amphibolic?
    (A) It occurs in both aerobic and anaerobic conditions.
    (B) It functions in both catabolism (energy production) and anabolism (precursor synthesis).
    (C) It produces both ATP and oxygen.
    (D) It operates in both prokaryotes and eukaryotes.

33. During fermentation, how is NAD^+ regenerated for glycolysis to continue?
    (A) By donating electrons to the electron transport chain
    (B) By reducing pyruvate to lactate or ethanol
    (C) By hydrolyzing ATP
    (D) By oxidizing $FADH_2$

34. Which of the following correctly describes the energy yield from one $FADH_2$ molecule in oxidative phosphorylation?
    (A) 1 ATP
    (B) 1.5 ATP
    (C) 2 ATP
    (D) 3 ATP

35. In wastewater treatment, what is the primary role of anabolic reactions?
    (A) To break down organic pollutants into CO_2 and H_2O
    (B) To synthesize new bacterial biomass using energy from catabolism
    (C) To reduce the pH of wastewater through acid production
    (D) To directly convert ammonia into nitrogen gas

36. When balancing a stoichiometric equation for bacterial growth in wastewater treatment, which of the following must be accounted for?
    (A) Only the organic substrate concentration
    (B) Electron acceptors (O_2), nutrients (N, P), and energy balance
    (C) Temperature and pressure of the wastewater
    (D) Color and turbidity of the effluent

37. Why is nitrogen (N) a critical component in anabolic reactions for wastewater bacteria?
    (A) It serves as the primary energy source for catabolism
    (B) It is required for synthesis of proteins and nucleic acids in new biomass
    (C) It acts as an electron acceptor in place of oxygen
    (D) It neutralizes toxic heavy metals in wastewater

38. How does the empirical formula of bacterial biomass ($C_5H_7O_2N$) assist in writing metabolic stoichiometric equations?
    (A) It predicts the exact species of bacteria present
    (B) It provides the molar ratios needed to balance elements in anabolic reactions
    (C) It determines the BOD_5 of the wastewater
    (D) It quantifies the dissolved oxygen demand

39. In a coupled catabolic-anabolic reaction for wastewater treatment, if the catabolic yield is 10 ATP per glucose molecule, what is the likely fate of this energy?
    (A) Entirely lost as heat due to entropy
    (B) Used to power anabolic reactions and cell maintenance
    (C) Stored indefinitely as glycogen
    (D) Released as light (bioluminescence)

40. Which phase of microbial growth is characterized by the highest metabolic activity and constant growth rate?
    (A) Lag phase
    (B) Exponential (log) phase
    (C) Stationary phase
    (D) Death phase

41. What does the Monod equation describe?

(A) The linear relationship between substrate concentration and growth rate
(B) The hyperbolic relationship between substrate concentration and specific growth rate
(C) The effect of temperature on enzyme activity
(D) The oxygen demand during fermentation

42. Which factor does NOT typically affect microbial reaction rates in wastewater treatment?

(A) Substrate concentration
(B) Dissolved oxygen
(C) Nutrient availability (N, P)
(D) Color of the wastewater

43. A bacterial culture has a generation time (doubling time) of 30 minutes. What is its specific growth rate (μ) in h^{-1}?

(A) $0.5\ h^{-1}$
(B) $1.0\ h^{-1}$
(C) $1.39\ h^{-1}$
(D) $2.0\ h^{-1}$

44. Under substrate-limited conditions, microbial growth rate is:

(A) Directly proportional to substrate concentration
(B) Independent of substrate concentration
(C) Determined solely by temperature
(D) Zero

45. The yield coefficient (Y) in microbial kinetics refers to:

(A) The ratio of biomass produced per unit substrate consumed
(B) The rate of oxygen consumption per unit biomass
(C) The fraction of substrate converted to CO_2
(D) The decay rate of cells in stationary phase

46. Given $\mu_{max} = 0.8\ h^{-1}$ and $K_s = 50$ mg/L, what is the specific growth rate (μ) at a substrate concentration of 200 mg/L?

(A) $0.16\ h^{-1}$
(B) $0.32\ h^{-1}$
(C) $0.64\ h^{-1}$
(D) $0.8\ h^{-1}$

47. In wastewater treatment, oxygen limitation typically leads to:

    (A) Increased growth rate due to fermentation
    (B) Shift from aerobic to anaerobic metabolism
    (C) Complete cessation of microbial activity
    (D) Higher biomass yield (Y)

48. The endogenous decay coefficient (K_d) accounts for:

    (A) Energy spent on cell maintenance and biomass loss
    (B) Substrate inhibition at high concentrations
    (C) Oxygen transfer efficiency
    (D) Nutrient uptake rates

49. Which bioreactor type maintains a continuous inflow and outflow of substrate while keeping biomass concentration constant?

    (A) Batch reactor
    (B) Fed-batch reactor
    (C) Chemostat
    (D) Plug-flow reactor

50. The Food-to-Microorganism ratio (F/M) is a critical parameter in bioreactor design because it:

    (A) Determines the color of the biomass
    (B) Balances substrate availability with microbial population to avoid overloading
    (C) Measures the dissolved oxygen concentration
    (D) Predicts the pH of the effluent

51. Which of the following units and formula combination is correctly used to calculate the F/M ratio in activated sludge systems?

    (A) mg/L ÷ MGD
    (B) (Q × BOD × 8.34) ÷ (MLSS × Volume × 8.34)
    (C) (Flow × BOD in mg/L) ÷ (Volume × MLSS)
    (D) (Q in MGD × BOD in mg/L × 8.34) ÷ (MLSS in mg/L × aeration tank volume in MG × 8.34)

52. In a completely mixed continuous-flow reactor (CSTR), the substrate concentration in the effluent is:

    (A) Higher than the influent due to biomass accumulation
    (B) Equal to the substrate concentration inside the reactor
    (C) Zero because all substrate is consumed
    (D) Unrelated to the hydraulic retention time

53. Which operational mode is characterized by no inflow or outflow during the reaction phase?

(A) Continuous stirred-tank reactor (CSTR)
(B) Batch reactor
(C) Chemostat
(D) Plug-flow reactor

54. A bioreactor has an influent flow rate of 10 m^3/day and a volume of 50 m^3. What is its hydraulic retention time (HRT)?

(A) 0.2 days
(B) 5 days
(C) 10 days
(D) 500 days

55. Sludge Retention Time (SRT) is a measure of:

(A) The average time biomass remains in the reactor
(B) The time taken for substrate to pass through the reactor
(C) The oxygen transfer efficiency
(D) The mixing intensity in a CSTR

56. A chemostat operates with a dilution rate (D) of 0.1 h^{-1} and a biomass yield (Y) of 0.5 g/g. If the influent substrate concentration (S_0) is 200 mg/L, what is the effluent substrate concentration (S) when $\mu = D$?

(A) 20 mg/L
(B) 50 mg/L
(C) 100 mg/L
(D) 200 mg/L

57. A bioreactor has an influent BOD of 300 mg/L and an effluent BOD of 30 mg/L. What is the BOD removal efficiency?

(A) 10%
(B) 70%
(C) 90%
(D) 100%

58. Which component of the activated sludge process is primarily responsible for biological oxidation of organic matter?

(A) Primary clarifier
(B) Aeration tank
(C) Secondary clarifier
(D) Chlorination unit

59. The Mean Cell Residence Time (MCRT) is a critical parameter because it:

(A) Measures the average time sludge remains in the system
(B) Determines the color of the effluent
(C) Calculates the oxygen transfer rate
(D) Predicts the pH of influent wastewater

60. What is the primary purpose of sludge recirculation from the secondary clarifier to the aeration tank?

(A) To increase influent flow rate
(B) To maintain active biomass in the system
(C) To reduce dissolved oxygen levels
(D) To eliminate the need for primary treatment

61. A high Food-to-Microorganism (F/M) ratio in the activated sludge process typically indicates:

(A) Insufficient substrate for biomass, leading to starvation
(B) Excessive organic load relative to microbial population
(C) Optimal nutrient balance for nitrification
(D) Complete removal of dissolved oxygen

62. Which parameter is directly calculated by dividing the mass of Mixed Liquor Suspended Solids (MLSS) by the daily sludge wastage rate?

(A) Hydraulic Retention Time (HRT)
(B) Sludge Volume Index (SVI)
(C) Mean Cell Residence Time (MCRT)
(D) Biochemical Oxygen Demand (BOD)

63. In biological nitrogen removal, nitrification requires which environmental condition to proceed efficiently?

(A) Strict anaerobic conditions
(B) Low pH (<6.0)
(C) Adequate dissolved oxygen and alkalinity
(D) High organic carbon concentration

64. The Modified Ludzack-Ettinger (MLE) process enhances nitrogen removal by:

(A) Combining pre-anoxic zones with aerobic nitrification
(B) Using chemical precipitants for phosphorus removal
(C) Eliminating the need for secondary clarifiers
(D) Operating at high F/M ratios (>0.5)

65. What is the primary electron donor required for denitrification?

(A) Oxygen (O_2)
(B) Carbon dioxide (CO_2)
(C) Organic carbon (BOD)
(D) Phosphates (PO_4^{3-})

66. A wastewater plant treats 2,000 m^3/day with an influent NH_4^+-N of 30 mg/L. If the nitrification rate is 1.2 g NH_4^+-N removed/g VSS-day and MLVSS is 2,000 mg/L, what aeration tank volume is needed for complete nitrification?

(A) 125 m^3
(B) 250 m^3
(C) 500 m^3
(D) 1,000 m^3

67. Biological phosphorus removal relies on the ability of PAOs (polyphosphate-accumulating organisms) to:

(A) Release phosphorus under anaerobic conditions and uptake it aerobically
(B) Convert phosphorus directly to N_2 gas
(C) Operate without organic carbon
(D) Function at pH > 9.0

68. What is the main function of phosphorus accumulating organisms (PAOs) under anaerobic conditions in the BPR process?

(A) They convert phosphate into nitrate
(B) They store phosphorus as polyhydroxybutyrate (PHB)
(C) They hydrolyze polyphosphate to store carbon as PHB
(D) They release nitrate and ammonia into the water

69. Post-anoxic denitrification differs from pre-anoxic denitrification by:

(A) Using endogenous carbon sources instead of influent BOD
(B) Eliminating the need for nitrification
(C) Operating at higher dissolved oxygen levels
(D) Removing phosphorus simultaneously

70. The PhoStrip process combines biological and chemical phosphorus removal by:

(A) Alternating anaerobic/aerobic phases with lime addition
(B) Using side-stream sludge fermentation to release phosphorus
(C) Bypassing primary treatment
(D) Increasing SRT to >30 days

71. What is the role of the anoxic zone in the PhoStrip process?

(A) To remove nitrogen by denitrification
(B) To promote aerobic digestion of organic matter
(C) To facilitate phosphorus uptake by PAOs
(D) To allow PAOs to release phosphorus for stripping

72. Which of the following is not an intermediate in the Citric Acid Cycle?

(A) Citrate
(B) Succinyl-CoA
(C) Acetyl-CoA
(D) Fumarate

73. Which structural feature of prokaryotic cells most significantly affects substrate uptake in wastewater treatment systems?

(A) Cell wall composition
(B) Flagellar arrangement
(C) Presence of nucleus
(D) Presence of mitochondria

74. Why are biofilms often more effective than planktonic cells in environmental engineering applications?

(A) They multiply faster
(B) They secrete more toxins
(C) They provide mass transfer limitations
(D) They form complex microbial communities with synergistic functions

75. Calculate the electron equivalents per mole of glucose ($C_6H_{12}O_6$):

(A) 12
(B) 24
(C) 48
(D) 6

76. Which of the following is true about bacterial energy yields under aerobic vs. anaerobic conditions?

(A) Anaerobic respiration yields more energy per mole substrate
(B) Aerobic respiration yields less ATP per mole of substrate
(C) Fermentation yields the most energy
(D) Aerobic respiration has the highest energy yield due to oxygen as terminal electron acceptor

77. If a wastewater has a BOD of 200 mg/L and a COD of 450 mg/L, what can you infer?

(A) The wastewater is highly biodegradable
(B) The wastewater contains a high fraction of non-biodegradable organics
(C) The BOD test overestimates oxygen demand
(D) There is no organic matter

78. Theoretical oxygen demand (ThOD) for a 50 mg/L glucose solution is approximately:

(A) 80 mg/L
(B) 100 mg/L
(C) 67 mg/L
(D) 90 mg/L

79. Which pathway is common to both aerobic and anaerobic respiration?

(A) Electron transport chain
(B) Fermentation
(C) Glycolysis
(D) Calvin cycle

80. Why is nitrate reduction critical in catabolism during wastewater treatment?

(A) It produces more ATP
(B) It enhances CO_2 fixation
(C) It serves as an alternative electron acceptor
(D) It increases sludge production

81. Anabolism in microorganisms primarily depends on:

(A) Substrate toxicity
(B) Availability of reducing equivalents and ATP
(C) Presence of toxins
(D) Dissolved oxygen only

82. Which macromolecule is most directly tied to cell yield in biomass production?

(A) Lipids
(B) DNA
(C) Proteins
(D) Carbohydrates

83. In Monod kinetics, if S << K, the rate of growth is:

(A) Zero
(B) First-order with respect to substrate
(C) Zero-order
(D) Constant

84. Determine the specific growth rate (μ) when $\mu_{max} = 0.6\ d^{-1}$, S = 20 mg/L, and K = 10 mg/L.

(A) $0.2\ d^{-1}$
(B) $0.4\ d^{-1}$
(C) $0.3\ d^{-1}$
(D) $0.1\ d^{-1}$

85. Which parameter is most critical to control biomass washout in a CSTR?

(A) pH
(B) Hydraulic retention time (HRT)
(C) Influent COD
(D) SRT (θ_x)

86. What is the purpose of recycle in activated sludge systems?

(A) Reduce oxygen demand
(B) Increase retention time of solids
(C) Increase flow rate
(D) Reduce nutrient input

87. Which of the following is not a common issue in activated sludge processes?

(A) Sludge bulking
(B) Foaming
(C) Nitrification
(D) Substrate inhibition

88. A bioreactor receives an influent BOD_5 of 400 mg/L. If the flow rate is 2,000 m^3/day and BOD removal is 92%, calculate the mass of BOD removed per day.

(A) 736 kg/day
(B) 368 kg/day
(C) 460 kg/day
(D) 920 kg/day

89. If you maintain an SRT of 10 days and yield coefficient $Y = 0.5$, what is biomass production per gram of BOD removed?

(A) 0.05 g
(B) 0.1 g
(C) 0.5 g
(D) 1.0 g

90. Enhanced biological phosphorus removal depends on which group of bacteria?

(A) Denitrifiers
(B) Methanogens
(C) PAOs (Polyphosphate Accumulating Organisms)
(D) SRBs (Sulfate Reducing Bacteria)

91. During denitrification, nitrate is converted to:

(A) Nitrite
(B) Ammonia
(C) Nitrogen gas
(D) Nitrous acid

92. In tertiary treatment, what is the typical role of alum or ferric chloride?

(A) Pathogen removal
(B) Nitrate reduction
(C) Phosphorus precipitation
(D) Sludge digestion

93. Which combination most accurately reflects an efficient BNR (Biological Nutrient Removal) system?

(A) Aerobic → Anoxic → Anaerobic
(B) Anaerobic → Anoxic → Aerobic
(C) Anoxic → Anaerobic → Aerobic
(D) Aerobic → Anaerobic → Anoxic

94. Why must alkalinity be supplemented in denitrifying systems?

(A) To buffer excess protons generated
(B) To aid biofilm formation
(C) To precipitate calcium
(D) To reduce ammonia

95. Calculate the net yield (Y_{net}) if $Y = 0.5, f_d = 0.8$, and $K_d = 0.06$ d^{-1}, $= SRT$ 6 d.

(A) 0.3
(B) 0.35
(C) 0.4
(D) 0.45

96. In a reactor designed for simultaneous nitrification and denitrification, how should DO be managed?

(A) Kept at saturation
(B) Maintained at zero
(C) Kept between 0.2–0.5 mg/L
(D) Alternated hourly

97. In biofilm reactors, substrate diffusion primarily limits:

(A) Cell yield
(B) Reaction stoichiometry
(C) Growth rate in interior layers
(D) Oxygen saturation in effluent

98. Which of the following accurately links biological stoichiometry to wastewater COD?

(A) 1 e^--equivalent releases 4 mg COD
(B) 1 g COD corresponds to 1 e^- eq
(C) 1 e^--equivalent equals 8 mg COD
(D) 1 mol O_2 equals 4 g COD

99. Which of the following best explains why denitrification can help offset alkalinity loss in nitrifying systems?

(A) It directly produces ammonia, which consumes protons
(B) It produces hydroxide ions during nitrate reduction
(C) It consumes nitrite which is alkaline
(D) It does not affect alkalinity balance at all

100. In designing a biological phosphorus removal system, the key operational factor to promote PAO activity over GAO (Glycogen Accumulating Organism) is:

(A) Higher temperatures
(B) Extended anaerobic contact with carbon source
(C) Elevated nitrate concentration
(D) High DO in anaerobic zone

MIX
Papier aus verantwortungsvollen Quellen
Paper from responsible sources
FSC® C105338

If you have any concerns about our products,
you can contact us on
ProductSafety@springernature.com

In case Publisher is established outside the EU,
the EU authorized representative is:
Springer Nature Customer Service Center GmbH
Europaplatz 3, 69115 Heidelberg, Germany

Printed by Libri Plureos GmbH
in Hamburg, Germany